STEW COHEN

iUniverse, Inc.
Bloomington

THE WYEN EXPERIENCE

Copyright © 2013 Stew Cohen

All rights reserved. No part of this book may be used or reproduced by any means, graphic, electronic, or mechanical, including photocopying, recording, taping or by any information storage retrieval system without the written permission of the publisher except in the case of brief quotations embodied in critical articles and reviews.

iUniverse books may be ordered through booksellers or by contacting:

iUniverse
1663 Liberty Drive
Bloomington, IN 47403
www.iuniverse.com
1-800-Authors (1-800-288-4677)

Because of the dynamic nature of the Internet, any Web addresses or links contained in this book may have changed since publication and may no longer be valid. The views expressed in this work are solely those of the author and do not necessarily reflect the views of the publisher, and the publisher hereby disclaims any responsibility for them.

Any people depicted in stock imagery provided by Thinkstock are models, and such images are being used for illustrative purposes only.

Certain stock imagery © Thinkstock.

ISBN: 978-1-4759-6961-0 (sc)
ISBN: 978-1-4759-6963-4 (hc)
ISBN: 978-1-4759-6962-7 (e)

Printed in the United States of America

iUniverse rev. date: 1/30/2013

I thank my parents for listening to me read the stories I wrote as a kid. They paid attention, regardless of the quality. My wife deserves credit for keeping the home repair list to herself. I also thank my sons, Brenden and Brant, for repeating, "Finish the book already!"

CONTENTS

Foreword .ix
Preface. xiii
Acknowledgments xvii

1. Radio Speak 1
2. From the Announcers to the Newsroom . . . 18
3. An Owner's Dream 28
4. Birth of Request Radio 52
5. Inaugural and Original Broadcasters 58
6. Home Sweet Home 75
7. Music Is My Friend 82
8. Tech No Savvy Seventies 90
9. Best Radio Reporter 95
10. He's No John Wayne 102
11. The Big Boys 111
12. Lessons Learned 119
13. News Is the Coolest 133
14. Could They Tap Dance? 152

15. Program Directors 174
16. Step in Tune with the Women of WYEN. . .179
17. Farella and Bryman. 191
18. American Airlines Crash 199
19. A Smile in Their Voices. 204
20. A Side Trip to Plains 211
21. Going About Their Business 217
22. Smooth Operators 225
23. The Everyman to the Marathon Man. 230
24. Keeping the Dream Alive. 247
25. There at the End 258

Conclusion: WYEN Memories 265

FOREWORD

Transcribing interviews is the most tedious job in radio. Fortunately, the digital world of today makes the process much easier than the analogue world of yesterday with reels of tape and cassettes. For my first documentary in 1975, I spent my entire winter break from college transcribing one hundred hours of interviews from dozens of cassettes. The documentary called *The Language Crisis* was only a half hour.

It is with an understanding of the amount of work required in transcribing that I appreciate Kenn Heinlein and his son, Dan. Kenn is a former WYEN announcer and son-in-law of the former owners of WYEN. He interviewed his mother-in-law, Carol Walters, for the foreword. Kenn gave Dan the audio interview to transcribe and write the foreword based on what his grandmother said about WYEN and her late husband, Ed Walters.

Chicago has always been a radio town. Look at a map, and it's easy to see why. Sure, plenty of people want to work in New York and Los Angeles, but on the coasts, even the strongest signals are either blocked by mountain ranges, wasted on the ocean, or both. But here, in the heart of the nation, where the skies are clear and the ground is flat, Chicago's most powerful radio stations carried for miles and miles. Kids would stay up at night listening to WLS and WCLF not only in Chicago, and not only throughout Illinois, but across the Midwest. When you did a show for Chicago, you were really doing a show for America. With that in mind, it should be clear why this wasn't the kind of place where a local broadcasting hopeful—someone just like Ed—would get right to work. In a way, it's a lot like being a baseball player working through the farm clubs. The neighborhood boy who did mock radio shows in his bedroom first might have to cut his teeth in backwaters too small even to name, doing anything and everything the station needed for mere pennies. Then, if he were lucky, that person might find work in midsized towns, such as Peoria, Grand Rapids, or Moline. From there, he could move up the ladder to something like Des Moines, Omaha, or Indianapolis, and if everything worked out, then maybe he could really make it and come back home to Chicago. Radio doesn't like to let you put down the stable roots that, say, refrigeration might, and that's burned many people out along the way. It's like the lyric from WKRP in Cincinnati: "Got kind of tired of packing and unpacking, town to town, up and down the dial."

So the radio road is long and hard. Ed and I learned this on Appalachian highways and Ottawa morning shows. We stuck it out, but who knows how many young people full of promise gave up along the way from having to toil in such thankless and unpromising work so far from home? We thought back to our own years of early struggling. So many people like us could have gone on to be like us now, until they realized they could make a better living doing almost anything but working sixteen-hour days in small-town radio. For someone who worked so hard for so long to live his dream, to give it all up at the first sign of adversity seemed unconscionable. Ed's next dream was to provide a training ground for the local kids just like him who wanted to dive right into one of the largest and most demanding radio markets, so they could learn the business and pay their dues without finding themselves disillusioned in some place like Watertown. As Ed gave this rare gift to the Chicago area, the area gave back with its brightest and hungriest aspiring broadcasters, engineers, and salesmen, who, realizing what an opportunity they had here, gave themselves to WYEN with all the selfless drive of any of the young hopefuls scrapping away on five-hundred-watt stations in towns of five-thousand people.

WYEN is long gone. The frequency has changed hands and formats more times than I would ever care to count, and if you drove past our old offices at Devon Avenue and Des Plaines River Road today, you'd never know what was there before. You can't find WYEN on the dial at 106.7, but you can find it in the people who have gone on, just like Ed wanted, to entertain millions of people in the most prized positions. People like Garry Meier, who has been a mainstay of Chicago radio for more than thirty years at WLUP, WLS, and now the station more associated with Chicago than any other, the venerable WGN. Or Bob Worthington, who went from spinning records for the northwest suburbs to broadcasting the Solid Gold Saturday Night oldies show for the entire country. Or Pat Foley, who called Chicago Blackhawks games over WYEN in 1981, and to this day remains the team's beloved voice. If Ed could see the lasting impact that he and his station have had on radio both in Chicago and beyond, he would be so very proud. Not only did so many broadcasters learn their trade under Ed, but those broadcasters went on to teach the next generation of radio talent, and that generation is doing the same today.

When we first began WYEN after years of hard work, we were proud to finally build our transmitter. Now, as I look back on my life in radio, I'm proudest of all that we built a lasting legacy of our hard work and long hours, made up of all the people who can trace their careers to two Polish kids from humble, old Bucktown who sacrificed immensely to live out a crazy dream. We hoped our tree would grow indefinitely, and it may yet continue. But I regret radio has changed a lot since our days in Des Plaines. The era of the mom-and-pop radio station has all but come to a close. Scan the dial today, and almost every station you find will be owned by some conglomerate, such as CBS, Bonneville, Citadel, Entercom, or Clear Channel. These groups will run as many as eight stations under one roof, and when they don't cut corners by filling a job at their country station with a jock from their sports station next door, they'll just turn the job over to some computer and dispense with hiring a person altogether. Chicago doesn't have its training grounds anymore, nor does any market, large or small. Even those old stations in the boonies are entirely automated today. I could never imagine this in our day; if one were to go off-air unexpectedly, there wouldn't be anyone around to turn it back on again. It's harder every day to find a station like WYEN, one built on a genuine passion for radio and the people in it, and it's even harder to find an Ed Walters, someone who worked hard, sharing his zeal with others so they could further share that zeal in turn. There may not be another.

Just as WYEN and everything it taught lives on through its alumni, now it will also live on through this fine written account, *The WYEN Experience*. It's a story that yearned to be told for years, and I hope you'll enjoy the recollections of Stew Cohen, Ray Smithers, Garry Meier, Rob Reynolds, Dave Alpert, Mike Tanner, Kenn Heinlein, Bob Roberts, and so many of the other branches of our WYEN tree. For this, their heartfelt tribute to the life, labor, legacy, and love of my husband, I am forever grateful.

PREFACE

Alice jumped into a hole and found Wonderland. She saw life underground firsthand and learned something about herself in the process. Although WYEN staff had to walk down a flight of stairs to the studios, we were entering a radio "Wonderland." Maybe I'm trying too hard for comparisons between Alice's adventures and the adventures of dozens of former WYEN broadcasters, but the stories of these talented people are told in *The WYEN Experience* from the inside looking out. The stories of the WYEN staff are similar in many ways to the thousands of broadcasters on radio stations across America in the 1970s and 1980s. Many of them hoped they'd land at the powerhouse stations with the call letters of WLS-Chicago, WGN-Chicago, KNX-Los Angeles, WSB-Atlanta, and KDKA-Pittsburgh. They'd put in their dues first and dream of the big time. Very few broadcasters, however, pushed past the secondary tier of radio stations. This level rarely drew much attention. I hope to show with this book on WYEN's staff, and their drive for success, that a secondary tier of radio stations deserves attention for developing talent just as much as WLS and WGN, the more publicity-generating magnets.

The book is aimed primarily at people interested in all things radio. Some are fascinated by the history of radio and the history of the events of an exciting decade, others enjoy reading about the talented broadcasters they've heard on-air, and still others remember WYEN and have a nostalgic place in their heart for Request Radio.

I make no real effort to convince young people to put down their iPods and iPads and read *The WYEN Experience*. They may not identify with

WYEN because the equipment in the 1970s and the music and society in general were so different from today. I understand how someone may not connect to an earlier time. I could not identify with the voices and music swinging out of the radios of the 1940s and 1950s. For a kid of the 1960s, I held a transistor radio to my ear as though I'd invented a new hearing aid and listened to The Monkees, Simon and Garfunkel, and the Beatles. Even though my parents danced to the Glenn Miller Orchestra and the Andrews Sisters of the war years, and my sister collected 45 rpm rock-and-roll records of Elvis in the late 1950s, I felt no great demand to learn anything about either era until school forced my hand to open up my history book with its chapters on American culture.

To successfully feel *The WYEN Experience*, I'm introducing the WYEN voices that were fully immersed in 1970s and early 1980s memories. Of course, the broadcasters behind "these voices" have moved on in their careers since the 1970s and 1980s, but more than thirty years ago, Greg Brown did not know that stepping through the door at WYEN would one day lead to WLS-FM.

As your chief guide into *The WYEN Experience*, I'll stand just outside the broadcast studio at my college campus radio station and wave you into my world. I'll lead the tour from college radio into the professional studios of WYEN where a leg dangles from the ceiling. The tour will walk over to the owner's office and meet Mr. Walters, the driving force behind WYEN Radio. We'll unveil the story together of how and why the Federal Communications Commission nearly broke him. Along the tour, you'll see the strategy of Ed's genius programmer and his surprising plan for WYEN. You'll step into an adjacent studio and view a young man with a severely sore throat trying his best not to completely lose his voice as he kicks off the inaugural broadcast. You'll get to learn the motivation of several young announcers who eventually reach legendary status. You'll be tempted with jelly-filled donuts brought to the station every day by a broadcaster whose family owned a bakery. The tour will also stop long enough to meet the sockless jock, the mountain man, and the every man.

On our tour, I'll have ample opportunities to embarrass myself in places where peppermint schnapps replaced good sense, where tennis great Bobby Riggs slammed me with Ping-Pong balls, and where Billy Carter didn't give me gas.

WYEN came early in my radio career from the end of the Vietnam War to the War in Iraq and Afghanistan. In those years, 1974 to 2012, I've won dozens of wire service awards from documentaries to features to reporting news. This includes news director titles at four radio stations. I've hosted job-shadowing programs, spoken at job fairs, and used my training to read to children in schools throughout Chicago's northwest suburbs.

WYEN breezed into existence in 1971 in Des Plaines, Illinois, flourished through the 1970s, stumbled in the early 1980s, and was sold a few years later. In the whole timetable of radio communications, WYEN represented one-tenth of the total life of radio history, a very small part indeed. Yet thousands of people listened to WYEN and enjoyed its music and announcers. The station mirrored the 1970s. I'm hoping through this book to give life to the station and the times.

Giving life again to the early days of great careers is also part of *The WYEN Experience*. For all of Garry's followers, imagine the interest in a young Meier on WYEN before his on-air partnership with Steve Dahl, or Alpert's unique news style on WYEN before Dave was heard through ABC Entertainment. These are radio gems before they were broadcast polished.

As promised, a little tour is in order. The scene is English class at Maine East High School. The year is 1971. It's early December, and I'm squirming at my desk because my handwritten speech is no better than chicken scratching on paper. I realize I'm in trouble. My confidence level is so low in both the text of my speech and ability to speak to anyone watching, I've got to make improvements.

"May I have a hall pass to the bathroom?"

ACKNOWLEDGMENTS

Three simple words are stuck in a perpetual life rewind and play. "Good luck, Stew." Ed Walters said these words the day I left WYEN. He had just let me go from the station, and I told him about my interest in the food industry. Laying out my future felt palatable rolling off my tongue.

Walters had owned WYEN. As general manager, he was not the type of man any of his employees could forget. His personality was so dynamic that we knew how his mind worked. That's why *The WYEN Experience* includes many people's recollection of his life. He felt so much at peace in the presence of talent. I interviewed many of his favorites for this book, yet despite all the words, one ingredient was lacking. Midway through research for the book, I received the special ingredient offering a depth of character that words could not adequately describe. Former WYEN announcer Frank Gray mentions in his interview with me that he had a rare Walters's aircheck of the brief time the boss jocked a WYEN shift. Gray made a copy and sent it. I pressed a button on my CD player, and Walters's voice played. Gray had sent me a dimension of Walters I thought was long gone.

The use of LinkedIn, Google, Facebook, and e-mails were instrumental in finding WYEN announcers that had scattered to Washington, Delaware, Wisconsin, Oklahoma, Florida, and other states. A few WYEN broadcasters seemed impossible to locate, even with periodic attempts over nearly three years of research and writing.

Thanks to my photographer, Robin Pendergrast of Robin F. Pendergrast

Photography, Inc. of Crystal Lake. Robin is one of the best portrait and landscape photographers in the United States.

Thanks to Kenn Heinlein for connecting me to the Walters family and introducing wonderful photos used in this book.

Thanks to Rob Reynolds and Ray Smithers, my go-to people for answering questions about WYEN's history.

Thanks to Jarett Reinwald for his critical approach to my book. His background in teaching English helped greatly.

Thanks to my parents, Sid and Shirley Cohen, for their support from the beginning of this project to the end.

Thanks to my wife, Rita, for listening to me read parts of the book to her and encouraging me to press on.

Illustration by Terry Sirrell, a cartoonist, humorous illustrator, and high school classmate of this author. Sirrell has been embedding his images in the American psyche for many years. You may have seen his work on the back of Cap'n Crunch and Kellogg's Corn Flakes cereal boxes. His cartoons and characters have also appeared in the advertising of major corporations, children's books and in dozens of publications. Check out his cartoons at www.tsirrell.com.

Chapter One

RADIO SPEAK

Giving a speech is the worst. Actually, speaking publicly falls somewhere between having a pimple growing in the middle of my forehead and tearing my pants at the crotch. I'm telling this to you from the boy's bathroom where I've just spent a half hour in a stall rewriting my speech. I'll come out, but the bell should ring in a few minutes. Of course my teacher knows I still have to give the speech. This is equal-opportunity suffering in high school English class, especially for those with performance anxiety. Science isn't quite as threatening. I find looking through a microscope rather comforting. Unfortunately, my advanced-placement high school biology grade won't warrant more attention to microscopes.

Why is it I remember all the high school experiences I'd rather forget and can't remember the experiences I wish I could recall?

My counselor, Miss Jane Simmons, waits for me in the hallway near my homeroom. I figure as much because she's standing in front of my locker. Simmons doesn't intentionally radiate an intimidating aura, she just has it naturally. Her facial muscles rarely form a smile, at least not that I've ever seen. Simmons could smell even a hint of fear tingling throughout a student's body. My throat tightens.

"Stewart, I need to talk to you," Simmons says, emphasizing my name with pinches of pity. "We must make a few changes in your class schedule."

"Okay," I choke out the word and look at her for a few seconds.

"Stewart, I won't let you take on more than you can handle."

Simmons opens her hand, realizing I am not her verbal sparring partner.

"Okay." I move my head up and down in an animated way, trying to hurry her along before a crowd gathers in front of my homeroom. Nothing worse than public humiliation—an introvert's living hell!

Just a year earlier, I realized I'd been shorted my Wonder Bread years and reluctantly let go of my most expressive and energetic dream of pitching the Cubs or White Sox to a World Series championship. All I had left in my field of view was the dream of becoming an author someday, but I realized quickly I couldn't major in Becoming an Author Someday. I'd have to take college-credit courses in English and possibly journalism.

Each of us has something special within us. Sometimes we find what that *something* is in high school, sometimes later in life, and sometimes never. I knew creative writing was my *something*. Stories I wrote and read into my Panasonic recorder with the joystick and little spools of tape recorded just fine in the confines of my bedroom alongside taping WLS and WCFL jocks introducing songs they played on their shows in the 1960s. Still, I never seriously thought of radio as a career because I had trouble with public speaking. But my love of all things radio never wavered or waned, just maintained under the personal baggage I carried in my head.

Richard Cohn, a childhood friend and high school classmate, was a member of the student-run WMTH (Maine Township High) and asked me in the parking lot at Maine East to join him on the radio station. Richard was fairly tall and lanky but had no problem sitting in my red 1966 Ford Fairlane 500. That car had lots of legroom. (Boys tend to remember their first car more than they remember just about anything else, except their first kiss.) I saw a lot of Richard in my homeroom and at his home. We hung around. Didn't really matter that radio might give us another thing to do together.

I could not imagine Harry Ford, Maine East class of 1960 (eventually to become Harrison Ford, the actor), had as much trouble deciding whether he should be part of WMTH. Former Maine East Principal David Barker told me the young Ford was one of the first student broadcasters on-air at WMTH. Richard and I didn't know anything about Harry *Harrison* Ford. He was still a little more than a year away from acting in a small role for the classic movie *American Graffiti* and five years away from fame in George Lucas's blockbuster *Star Wars*. Even with insight into Ford's future, the knowledge of his career path wouldn't have drawn me into the radio station. I mean, Richard implied I'd have to talk on the radio. *Talk on radio!* How? Public speaking paralyzed my mouth and squeezed beads of sweat from my hands.

"Radio is for girls!" I blurted.

I shot down his fairly innocuous suggestion with my own controversial statement that could not be defended by any rational person.

THE WYEN EXPERIENCE

"Okay," Richard let out, opening his eyes wider, turning his head, looking out the window.

Normally, Richard strung more words together in actual sentences that might not end unless I'd interrupt. That's why Richard was good for radio. He'd fill all the spaces where dead air might lurk.

My words were unsettling, rattling around my brain, alerting some inactive neurons. An internal fight was brewing. But I'd been too fearful to take back what I said … and so I paid a price for my fear of public speaking. That fight had to wait for another time, certainly not our visit to Chicago and WCFL.

Before college, the closest I came to visiting a radio station was in 1967. Richard and I rode the Skokie Swift, transferred onto the El, and spent a Saturday downtown. We meant to see WLS-AM and WCFL-AM. We made it over to the Marina City Complex at 300 North State Street and found the building housing WCFL. In 1967, WCFL promoted itself as Big 10 WCFL, Chicago's number-one contemporary radio station. WLS and WCFL were among the most popular music stations in the United States, and the two were competing in radio wars. This was certainly an exciting time for a visit. We were thirteen years old in the summer of 1967, just a few months shy of eighth grade. The radio stations were everything to us in the summer. I switched back and forth between WLS and WCFL, listening to the jocks, taping the stations' music on my recorder. I'd sing and talk into the little microphone and listen back until I bored myself silly. We were radio geeks following the light of broadcasting. I couldn't wait to see one of the jocks—maybe Barney Pip or Jim Stagg or Ron Riley or Joel Sebastian at WCFL.

We walked to the main floor of what we believed was the radio station and stood in a large room. In a corner of the lobby, a woman sat behind a desk, wearing a phone in her ear and pearls around her neck. She answered calls over a steady hum, punctuated occasionally by horns or tires squealing from braking cars, taxis, and buses outside. The scene I had pictured on our walk from the train station to WCFL immediately disappeared. No announcer talking on-air, sitting behind a ton of equipment—speakers, a sound-mixing board, and turntables. No huge metal microphone almost in his mouth. No gesturing with his hands, pointing occasionally while others around him, mostly young guys in white shirts with black ties, scurried here and there, bringing copy, records, black coffee, and a clean ashtray. No distinctive beat of "Kind of a Drag" by The Buckinghams, "I'm a Believer" by The Monkees, or "Windy" by The Association played in succession, except for the moments the announcer snuck in a few wisecracks between songs.

"Stewart, I don't see anything in here," Richard whispered.

"No tours on Saturdays," the female voice instructed, waving what looked like orange bumper stickers.

I whispered back to Richard, "We came all the way downtown."

"I have stickers and pictures of some of our announcers," she said, holding stickers in one hand and pictures in the other. "That's the best I can—"

"Excuse me, but where are the announcers?" Richard interrupted.

"Unfortunately, an announcer won't pop out from behind a wall," the receptionist said. "Please take some of these."

We put our hands out.

"Thank you," Richard said quietly, looking dejected as though he either needed to find a bathroom quickly or had an abscessed tooth.

We walked out. Well, not exactly. Richard walked out. I walked backward for a long, last look. Good thing. Although I realized there wasn't any radio stuff in the lobby, I saw the WCFL stickers I'd been given laying on a table where I had set them down. I think the light in the lobby hit them just right. There was this orange glow I couldn't miss.

"How about we go over to WLS?" Richard asked.

"Do you know where it is?"

"Yeah," Richard said.

Richard could be rather convincing, but this wasn't one of those times. I could not see myself walking all over downtown Chicago, possibly getting lost, so I said, "Let's just skip WLS."

Richard nodded.

Legendary announcer Clark Weber told me years later the WCFL viewing area was not designed for much viewing.

"It's a shame you and Richard didn't walk over to the WLS Studio at Michigan and Wacker. We had so many kids coming in on Saturday that we had an Andy Frain usher moving the kids out every fifteen to twenty minutes so we could make room for another bunch of eager kids. The only drawback to the Saturday viewing at WLS was that the unsocial Mr. Lujack would pull the drapes because he said the snotty-nosed kids were a distraction, which they may have been, but Larry's social graces were never his strong suit!"

Weber hosted the morning show for six of his nine years at WLS-AM, 1961 to 1970. He served as program director for two years. Clark also worked for WCFL, WMAQ, WIND, WJJD, and WAIT. He's currently president of Clark Weber Associates, an advertising agency.

Along came *Washington Post* journalists Bob Woodward and Carl Bernstein, bringing down a president and his closest advisors. In the summer of 1972 and over the next couple of years with the ongoing investigation, I gravitated into the journalism program at Southern Illinois University. Some students were star-struck by Robert Redford and Dustin Hoffman

THE WYEN EXPERIENCE

playing Woodward and Bernstein in *All the President's Men*. Others didn't need inspiration from the massive aura of a Redford or Hoffman. Those young journalists believed they could make a difference in the world with their writing talent. Early in my career, a few people thought I looked like Hoffman, but no, that wasn't the reason I chose journalism. I could put my thoughts together on paper much better than I could say what I was thinking. I thought I'd study journalism in college, preparing for a position on a newspaper as a journalist capable of writing stories from school board to features. I'd gladly leave the taking down of presidential administrations to the big guys of the *Washington Post* and *New York Times*.

Had I written this part of my book for Hollywood, I'd say fate stepped in. A bit dramatic, maybe, but being at the right place at the right time sure fit. The news director and assistant news director of the campus radio station, WIDB, were living on the seventh floor of Schneider Hall on the Southern Illinois University campus in Carbondale, Illinois. My room was on the same floor, but down another hallway. What were the odds? Schneider Hall had more than a thousand male students in the sixteen-floor dorm. In my sophomore year at Schneider Hall, I made friends with the campus radio station Assistant News Director Bob Comstock. Comstock shared a room with the campus station's News Director Don Strom. Between Don and Bob, I'd hear the goings-on of staff at the radio station, and what I heard sounded like fun. They talked of news events, hashed over radio day-to-day stuff, and gathered informally at Pinch Penny Pub (a great pizza and beer place just off campus.) On weekends, Bob worked an afternoon news shift at the radio station. So he'd excuse himself and head over from Schneider Hall to Wright I where WIDB was housed in the basement.

Bob's easygoing personality mirrored one of the popular TV characters of the day—Bob Newhart playing Dr. Robert Hartley. Bob was funny without trying hard. He just had a great sense of timing. He did one of those double takes that made *Tonight Show* host Johnny Carson such a great entertainer.

For a few weekends in a row, we practiced this ritual of meeting in his dorm room. We talked about radio, school, and our weekend plans. We'd take the elevator down to the lobby and stand outside for a few minutes in front of Schneider Hall. He'd head off to WIDB. I don't know exactly what changed, but I got an epiphany. *This time, I'm tagging along.*

"Bob, wait! Can I go with you?"

He stopped, turned, and walked back to me.

"Kind of boring, you know, for someone sitting around in the station," Bob cautioned.

"That's okay," I assured him. "It's got to be better than sitting with the guys in the dorm lobby watching whatever garbage they've put on—"

"Come on," he interrupted, pointing at the small, black WIDB sign. "I have a newscast in a half hour."

"Didn't realize the station is as close as it is," I admitted.

"Hey, I've got the door," Bob said.

He opened the thick metal door blocking the music from floating out onto the campus.

"Oh," I exhaled, stepping into a radio station for the first time ever.

"Find a place to sit. It'll be a while," Bob said.

"That's 'Seasons in the Sun,'" I parroted the announcer introducing Terry Jacks's big hit, and then lip-synced the song. The music lounged through the air, circulated into the lobby, and filled the production studio before bouncing off the metal door.

"Give me a few minutes. I'll get back to you, Stewart," Bob said from somewhere off in another room.

"Seasons in the Sun" dimmed into Elton John's "Crocodile Rock." The jock carefully spun the cart rack. Carts with spots, jingles, songs, and jock promos glided past. The pounding beat on the speakers from the song on the turntable brought energy to the room.

From where I sat, I saw another song pulled from the rack. I moved my chair away from the wire service machine and closer to the newsroom.

Bob returned. "I just have to do this newscast, and we can go."

"Good," I answered, trying hard to keep my excitement level down, because I was now in the radio station and sitting in the best seat.

Bob prepared in both the newsroom and on-air news booth. I looked around where I sat just outside the newsroom. While the newsroom had a rectangular table, chairs, and a typewriter, the main room where I sat had a water fountain, clock, wire service machine, and speaker. I watched Bob rip wire copy from a very loud, clacking United Press International machine. He retyped news stories and moved to the on-air booth where he adjusted the microphone and headphones, cleared his voice, and closed the door in this room half the size of a walk-in closet. Bob pressed a button that switched on the on-air light in the room. He waited for the jock's cue. Thick glass panes separated the studios, so however forcefully Bob cleared his throat, the sound wouldn't go on-air in the jock's studio if the microphone was accidentally left on. Bob looked calm; he didn't move around in his chair or reposition the microphone or do anything that might suggest nerves. This was the first time I'd see anyone perform on radio.

The announcer's index finger chops down through the air. Instantly, Bob lowers his neck, hunches forward, and straightens his shoulders. He begins reading.

THE WYEN EXPERIENCE

Bob Comstock reads through a story on Vietnam in the WIDB News booth.

"President Nixon today discusses with Secretary of State Henry Kissinger the number of soldiers killed in the latest round of fighting in North Vietnam," Bob's deep authoritative voice pausing between the words *soldiers* and *killed*. He groups the words *latest round of fighting* with an inflection on *fighting*. Even with a higher tone he uses at the end of the sentence, Comstock's voice comes out richer than on the radio in my dorm room.

In the main room, stories are spewing from a box of yellow paper fed through the UPI machine. The machine with all the news is shaking side to side and forward and backward as though it's washing a load of clothes on spin cycle. Bob is seeing none of this. However, hearing him is more difficult with all the mechanical activity.

With even pacing, Comstock reads through his copy from start to finish.

7

He's not falling into the trap of a lot of college radio newscasters, starting slowly, gaining speed, and finishing as if they are in a race with Triple Crown winner Secretariat.

The lower register of his voice lets out a buzzing sound. Words fly every which way. Bob seems wary of the amount of copy left as he nears the last page of his news. I learned later he had pretimed sports, weather, and the tag he used for ending the newscast. He has thirty seconds of copy, right at the 4:30 mark on his stopwatch from reading the body of his newscast.

Today, tonight, tomorrow are covered and the sky conditions.

"I'm Bob Comstock, WIDB News."

Absolutely impressed by Comstock's performance, I thought I was listening to a Chicago radio newscast.

For just that one moment, he made it look easy for anyone off the street to sit in the booth and read news. Maybe I could try? You believe hard enough, you could sell yourself on anything, but this *bill of goods* doesn't come with a receipt.

The full feel of music once again pounded out of a speaker secured to the wall above the wire service. "Dream On" by Aerosmith vibrated the grill until lead singer Steven Tyler's voice trailed off and the grill stopped vibrating. The announcer didn't wait until the song ended. He walked around the WIDB studio picking out more music.

"Stewart, I'll be with you shortly," Bob shouted above Tyler. "I've got to get those stories initialed and dated and pegged on the wall."

He'd taken a quick peek outside the newsroom.

Sounding the least bit pushy, I asked, "Is it okay if I sit in the news booth?"

"Sure, you can go in there now."

Bob had left the light on. Some of his copy rested on the desk, some of the copy on the floor. I shut the door, and unexpectedly, the music I'd heard earlier played muted in the booth.

Maybe this isn't soundproof.

"Testing," I repeated, "testing, testing, testing."

I grabbed the microphone in a fist and sat down in Comstock's chair, staring at my reflection in the glass. Seemed my face inadvertently crinkled. This look surfaced again at WYEN after someone sprayed a heavy dose of disinfectant on the microphone wind screen. Then I had more trouble gagging from the spray dripping off the microphone into my lap.

Bob noticed my contorted face.

"Is everything okay?"

The closer I moved the microphone to my mouth, the more uncomfortable I felt.

THE WYEN EXPERIENCE

"No problems," though lying to Bob didn't feel so good. I could not ignore my problems with public speaking. Yet, I sensed a new direction.

The summer before my junior year, I thought of changing my major from journalism to radio-TV. However, this was somewhat risky. I hadn't yet found out whether managing my fear was possible. Although I'd been Bob's guest at WIDB six months earlier and decided to audition for radio news, I played it safe by sticking with a major in journalism.

Back on the Carbondale campus, WIDB staff announced auditions. My *radio is for girls* excuse from high school was something I'd never say again. Muttering that totally feeble and incorrect assessment of the radio industry would not factor into the latest excuses: a sore throat, bloodshot eyes, too many college research papers, or a hangover. But those excuses only masked a fear of public speaking, and I surely felt I could control my problems.

Bob was at WIDB. I called and told him I'd drop by.

He was preparing for a newscast, going through the steps he'd showed me before.

"Are you and Don still holding auditions?"

Bob put down his news copy. "I wondered when you'd get around to it."

He didn't need to know all the thinking I'd done over the summer and my hesitation because I didn't want to embarrass myself.

"We've just started the auditions. I'll put your name on the list."

"Thanks," I dragged out the word, believing a quick thanks wasn't good enough.

The day came. I looked forward to the audition one moment and dreaded it the next. At the radio station, several news hopefuls waited in the lobby. Some sat on chairs. A few hopped up on a table and sat. No one looked particularly nervous. A few had trouble keeping their hands still or a foot from tapping on the floor, but they were mostly quiet. One guy—that would be me—was clearing his throat every few seconds. Don came by, smiling and looking very calm.

"Stewart, would you like some water?"

"Sure, Don. Some water might help."

He came back with a cup of water a moment later.

"Thanks." I took the cup, but I couldn't tell him something stronger might be better. My little throat-clearing habit was more of a nervous thing than anything else.

One by one, each of the people there for an audition left the lobby as Don led them to another room. Suddenly the door to the station exit looked inviting, but I had already waited over an hour. Just then, Don and Bob walked over. They directed my attention to a cubicle, where they set up a Wollensak 3M reel-to-reel tape recorder and microphone.

"Press the record button and play button together, and then start reading your news when you're ready." Strom's instructions seemed pretty straightforward.

"I'll wait a few seconds and push through this," I mumbled. Checking the copy no less than a dozen times, I put every piece of paper in order of importance and then reached for the record button, but stopped abruptly. Was anyone listening in any of the nearby cubicles? I couldn't trust the silence. A chair in my cubicle made a pretty good step stool. Things looked okay. It should not have mattered whether five or twenty people were listening and watching. Confidence though was in short supply.

Might not seem like much to think about, but holding wire service news copy in front of my face and then picking up another piece of copy with a news story typed on it takes a bit of coordination and practice. What do I do with the read copy? Just put it back on the table, copy side down, or let the paper fall to the floor? We don't worry about this today because the computer acts as a teleprompter. The copy is on the computer monitor screen. Clicking on the screen's arrow scrolls down the page—at least that's the way I read news. But just in case of a power outage, stations without backup generator power should not lose their saved script.

Each piece of copy floated to the floor. An indoor snowstorm of news stories buried my feet. Before trying to hear the newscast, I shoveled the stories off the floor, put them back in order, and tried to play the newscast, but nothing came out of the recorder.

They'll think I know nothing about radio and broadcast equipment. Maybe they're right.

Don and Bob did not judge after I brought Don back and described my problem with the taping. Don smiled and put me through the whole instruction thing again.

The words *stinking raw* adequately described my first audition. The news director turned down my attempt, and if you had heard my audition, you'd agree. He was totally justified. My audition stunk. He and I knew it, but I held out hope that his *sense of smell* wasn't working.

"You did not make it *this time*." Strom let me down as gently as he could, but he encouraged another try.

I'd have asked him where he thought improvement was needed, but the truth is often a bit scary. My skin wasn't yet tough enough, so I only asked, "When is your next audition?"

Strom's opinion had value, not just as news director, but as a predictor of talent. But losing even the little amount of confidence I had left might make me give up. Why couldn't I just have a modicum of Strom's talent? Strom was perhaps the best candidate I'd seen for a TV talk show host. He had a pleasant yet powerful voice and delivery, good looks for television, and stood just over six feet tall. I figured he'd either host a news show on television or anchor television news.

By my fourth audition, he said the same thing as the earlier auditions. Nothing changed from his end, but I sensed improvement in the third and fourth auditions.

What does it take to sound conversational? I wondered, repeating this in my head, walking to my classes on the SIU campus.

I TOOK A BREAK FROM STUDYING AT MY APARTMENT IN LEWIS PARK.

Did I think someone would unlock the methods for a successful newscast? The answers weren't forthcoming! Then I realized I went about this wrong. Instead of improving by some form of osmosis, try reading anything and everything out loud. Well, I exaggerated about reading *everything*. Read to anyone who wouldn't mind listening. Get feedback. Jeff Brenner might hold the solution. We'd been sharing an apartment in the newly built Lewis Park just off the SIU campus. It didn't bother me in the least asking Jeff for help. That's what best friends are for, and he'd been good at being the friend I could trust more than any other.

Besides Jeff, two other guys lived in the apartment, and we all studied in the evening. I read so loudly they may have been justified to tape my mouth shut. Someone knocked on my door.

I yawned, clearing my head. "Come on in."

Jeff walked in.

"The guys can hear you reading, and they can't study."

One thing about Jeff, he could totally fool me. He had this great poker face, never expressing even the hint of a smile, just a serious look with his eyes slightly narrowed. After so many times where he'd pull one over on me, I started to doubt his sincerity. He'd eventually break character if I did not react.

"Stewart, you're too loud."

"I didn't know the doors and walls are so thin, Jeff."

"Your voice pierces through walls."

He wasn't giving up. I conceded; maybe my loud reading was a distraction.

"Yeah, I could tone it down a bit."

He turned to leave.

"Listen, Jeff, while you're here, I could use your help."

"What?"

He turned back to me, pulled a chair from my desk, rolled it over to the bed and sat.

I explained, "For the past few days, my English Literature class assigned a lot of short stories and reports on each. I've fallen asleep on the bed and couch, and sometimes on the dining room table. Could you occasionally listen in and give me a few elbows … and critiques?"

Jeff hardly ever took time digesting stuff. He was quick with answers, and this time was no different.

"I'm thinking of a tradeoff."

"What do you mean a tradeoff?"

"I'll let you read to me … but I'd like your opinion on my architectural renderings."

"You mean you'd like me to tell you what I think of the work you do for your Interior Design classes?"

"You know, I'll show you a dozen chairs, and you just let me know the one you like best, or I'll show you the one I picked and tell you about it, like the Eames Lounge Chair. You'd see renderings on an interior of a bank, the lobby of a hotel, or the dining room of a restaurant I've designed.

"Jeff, all of your work is excellent! I can't criticize any of it."

We helped each other, only I good-naturedly endured his occasional attempt at transforming himself into Ted Baxter on *The Mary Tyler Moore Show*.

"Stewart Cohen, CBS News."

"Jeff, that's a caricature of television and radio anchors; it's exaggerated … come on!" But, Jeff doing Ted Knight doing Ted Baxter was really pretty funny because Jeff put his chin to his chest and then got the sound from all the way down in his shoes … rolling back up through his chest and straight out of his mouth. I thought he had grabbed a toilet paper tube and talked through it.

For several months, I read everything out loud, whether it was homework or a novel. I just read, read, read. Next time Strom called an audition, he'd get blown away with my "new delivery." Unfortunately, my voice became hoarse before the scheduled fifth audition, and while some people might just pack it in, I figured a sixth audition was my best shot. I did not judge the "sound" so much as the "feel," and I felt more confident. In what I hoped was my final audition, I waited outside the main studio for the news director. He brought news copy and took me over to a cubicle for another encounter with a reel-to-reel recorder.

"You know what to do with the copy, Stewart," Don said. "I'll leave the room in a second."

"I've practiced a lot, Don."

He was already turned to go but heard me and turned again, opening up his hands, nodding as though he knew I was ready this time.

"Really nothing to worry about, Stewart. You don't make it this way, and you've declared your major in radio and television, the department will have you on WSIU-FM doing news or sports. You make this audition, and you're on WIDB and WSIU."

The copy went in order of national and international first and second, followed by Illinois and local news, and then sports and weather. I laid down my news track and almost liked the playback. Don must have known. He walked into the room just as I turned off the playback and hit rewind.

"Don, I'm done."

He reached out with his hand, and I placed the reel in his hand, but not without a momentary hitch. I almost didn't let go of the tape.

"I'll let you know, Stewart," Don said, leaving the room with my audition.

So many other times, he said the same thing. How was this time different? I left WIDB.

Hours later, he called. I held my breath, not unlike other really nervous moments.

"I've got good news for you," Strom said. "Are you doing anything Friday and Saturday nights?"

Don offered an opportunity. His question didn't fool me, yet every word he spoke replayed in my mind. I'd waited too long to hear these words, and I was darn ready to give Strom's words a real workout in my mind. I had practiced for just this moment, dreaming what I'd say … no, walking the campus and saying what I'd say. I think I spent a lot of time imagining him asking me to do the news, and I'd been disappointed time and again by reality, until …

"I'm just studying," I said, answering his question.

"Your shift is 10 p.m. to midnight, Friday and Saturday," Strom asserted.

"I can't believe it! Thank you so much." Frustration over too many failures stopped me from welling up with tears. I had gone through a self-imposed, painful self-examination for months, in which I came to the understanding that reading news was one of the hardest things imaginable, and that no one had an exact answer to the mystery of sounding conversational.

Strom's career successes were easy to follow. He used his fine-tuned ability as a communicator for law enforcement, landing with the Carbondale Police, eventually becoming police chief, deputy director with the Illinois Secretary of State Police, and inspector general. Today, Strom serves as police chief for Washington University, St. Louis.

Comstock's even-tempered demeanor and ability to see humor in most things made him a natural for education. Bob is on staff at Homewood Flossmoor High School and is serving as general manager of the high school's broadcasting program.

By Monday morning, the euphoria of Don's announcement slid into apprehension. My first broadcast was only four days away, and with each day passing, I felt as though I walked on icy sidewalk squares, tentatively moving forward a few inches and then sliding back a bit.

The reality did not look promising. No amount of imagination could compensate for a huge lack of confidence at the prospect of laying my reading skills bare for all to hear. First I tried the imagination route. I dreamed of anchoring network news. My broadcast idol, Reid Collins of CBS Radio, suddenly was me. TV news giants Walter Cronkite and Frank Reynolds were my equals. A couple of days of this helped.

Being a creature of many nervous habits, I did everything the same way most days. Why jinx myself by doing something different? Friday started moving faster in the afternoon ... zooming forward within the last few hours leading to the 10 p.m. to midnight shift. I considered drinking beer but recognized this was a crazy thought.

The imagination route got me to Friday, but the nerves kicked in something fierce, and I couldn't eat or think of anything beyond reading the news. I tried my second route. Some people turn to God in moments such as this. They're in need of inspiration and divine guidance. I prayed because I needed someone or something stronger than myself sitting in the newsroom with me, guiding me through.

There must have been a reason why Tony Waitekus stood in front of the jock studio. He just popped up. Maybe he'd been working in a separate production studio. Mine wasn't to question, just to be grateful the WIDB news anchor was available if I lost the tug of war going on in my head.

I can't ask him to read my news. I'll go through with this. Maybe I should ask him, but that's it, I'll be done with this.

Logic wasn't winning the imaginary tug of war.

I have to do this! Maybe I could ask him for a bottle of Jack Daniels? Ridiculous! Why would he possibly carry Jack?

Tony's voice was my lifeline. "I see you've stopped pacing. That's all normal, you know."

"Does this get any easier, Tony?"

"Stewart, you'll be just fine," Tony calmly asserted.

I pressed, "How do you do it?"

"Read news?"

"No," I said, "how can you stay calm before you go on?"

"I am nervous ... you just won't know it."

Waitekus looked down momentarily, adjusting his glasses.

A professor once suggested picturing a calm lake in a forest. I thought of Thompson Lake in Thompson Woods at SIU. This was the perfect place on campus where the water is as still as glass. Trees along the edges offer shade for the tall grass growing unabated. Not even a ripple disturbed the surface reflecting white, puffy clouds.

"Nervousness can't be stopped, nor should it, Stewart," Tony explained. "However, you can control your nerves ... make it work for you."

He could hear and see my panic. If faster breathing, fidgeting, and throat clearing weren't enough, my eyes were bulging too.

"You will live through this, and you'll do another newscast and then another," Tony predicted.

I was lucky to have Tony at WIDB that evening. He told me he got through his first on-air experience without suffering a nervous breakdown.

He joked. Tony probably wondered at what time he became a psychologist to the radio-challenged.

No doubt Waitekus has helped many broadcasters starting out in the business. His career has taken him from Carbondale at WCIL AM/FM, where he served as music director and program director, to WIFC Wausau, Wisconsin, as program director. In between, Waitekus held titles as music director and program director at WTAO/Murphysboro, Illinois, and program director at WHTS/Rock Island Quad Cities. He also worked in radio in North Carolina and Green Bay, Wisconsin.

After Tony left the room, I repeated, "Who's nervous? I can do it. Just follow Strom's advice for the newscast."

I closed my eyes, but the light of the newsroom on the inside of my eyelids didn't give me the darkness of calm.

On went the headphones, on went the on-air light, and on went the newsroom microphone. Tom Sheldon cued me from the adjacent deejay studio. Gulping down my saliva, I opened my mouth, and set off an even faster heartbeat.

"I'm Stewart Cohen of WIDB News ... and in the news today ... President Nixon ..."

Slightly unsteady, hesitating between sentences, my voice gave the impression of a boat motor refusing to fully turn over. The boat never glided across the water. Declarative sentences sputtered; the boat took on water. I welled up with saliva and slowed my delivery because the lack of air choked off the engine.

I can't remember the rest of the newscast, but the cast sunk quickly.

Painfully, the whole cast lacked style, as I had heard it on WLS and WBBM. Something came over me: fear. Something else quickly appeared: loss of air. The two together made a unique combination: hyperventilation. I didn't know what hyperventilating was all about, but in this tiny newsroom of WIDB, my body seemed to start shutting down. Five minutes of news seemed equivalent to an Olympic-size marathon. Only twenty seconds passed, and I couldn't get air into my lungs. I tried deep breaths after nearly every story but could only bring in small amounts of air because I just didn't have time to sit there sucking in air. A dizzy, lightheaded feeling came over me a minute into the newscast, like waking up with a hangover. I sucked in air, read for a bit, sucked in more air.

"In Carbondale, it's fifty-five. I'm Stewart Cohen, WIDB News."

That's how it's supposed to end, just like that ... nicely spaced, even tempo, but my delivery ran the words together:

"In Carbondale, it's fifty-five ... (sucking air) ... I'mStewartCohenWID BNewwwwsssss." (All the air went out of the balloon!)

"Good job!" Sheldon yelled through the glass.

Some lies are helpful, some are hurtful. His was helpful. I sure didn't fault him for it.

A Chicago native, Sheldon, spent much of his professional life in downstate Illinois as an announcer for both WDML and WMIX in Mt. Vernon. After he graduated from Southern Illinois University in Carbondale in 1978, Tom worked at KGMO in Cape Girardeau, Missouri. Sheldon died at the age of fifty-three in December of 2008. In a published report in the *Mt. Vernon Register-News*, Sheldon was described as possessing a "golden throat."

WLS News Director Bud Miller visited the SIU Carbondale campus in 1976. Miller met with student broadcasters, gathering in the University Center auditorium, a large room with plush seating for all kinds of events. Radio-TV students and young broadcasters at WIDB and WSIU packed the auditorium for a session on the profession of radio. Someone turned off most of the lights before Miller began speaking. I nodded off shortly after he started talking and probably started snoring too. My choice of seats wasn't the best for someone sleeping during a lecture. Miller could see several rows back from the stage, and in the middle of the auditorium, he saw a head down. Not exactly the impression I envisioned.

"You have to have energy to wake up at 3 a.m. and be at work by 4 a.m. for your shift. You can't sleep like this guy in the audience," Miller admonished, but fortunately, he didn't point me out. The room immediately quieted, and I woke up abruptly. No one in the audience figured out which of their classmates was sound asleep, and Miller wouldn't embarrass me further. Years later, I appeared on a panel with Miller and told him about the incident. He laughed and asked me if I've managed to get enough sleep since then.

Two years after I nearly passed out reading WIDB news, Walters hired me as his evening news anchor and daytime reporter. He opened the door for someone with literally zero broadcast experience. He knew my level of experience and was willing to gamble.

CHAPTER TWO

FROM THE ANNOUNCERS TO THE NEWSROOM

Many of the broadcasters you'll meet in *The WYEN Experience* listened to great radio as children of the sixties. The Chicago stations of WLS and WCFL were home to the gold standard of announcers and the music they played. We were so influenced by the exuberance of Chicago radio announcers, we imitated those deejays. We introduced records and singers from songs we'd seen listed that week on the WLS Silver Dollar Survey, read ads from *Time* and *Boy's Life*, and tossed in humor if we were funny or were allowed to stay up late and write down one of Carson's jokes in his *Tonight Show* monologue. We also read news from the *Chicago Sun-Times*, *Chicago Tribune*, *Chicago Daily News*, and *Chicago's American* (later to become the *Chicago Today* before folding.)

"I'm Stewart Cohen, and this is your WLS News."

Our high-pitched voices struggled for a lower register.

"'To Sir with Love' on The Big 89, WLS."

Unfortunately, our voices hadn't changed, though we'd begun the voice-cracking stage. We were just a shave away from free-flowing testosterone, enough for growing peach fuzz on our faces. Then we turned up the volume on our tape recorder, playing back the show we had taped earlier. Lulu sang

THE WYEN EXPERIENCE

her hit song from the movie. Then we hit the button labeled *stop* on the recorder, and both posts spinning the little reels stopped.

Our little performances impressed no one because this was our secret fun, our fantasy of thinking cool and introducing music. We were boys and girls in the summertime of 1965, 1966, and 1967 playing with our transistor radio and recorder with the door closed. We'd die of embarrassment lest someone walk in on us and ask to hear what we're saying or hear us singing into the microphone, or playing back our off-key vocal of such popular songs as "Wild Thing" by The Troggs or the girls singing along with Bobby Sherman. The next day, we taped more WLS and WCFL music—The Buckinghams, Herman's Hermits, and Petula Clark—and again we said things that might sound goofy to our older brothers and sisters or our parents, but we never let them in on our deejay show.

Bill Jurek and Nick Farella were certainly not alone listening to and idolizing the work of Clark Weber, Art Roberts, Dick Biondi, Ron Riley, Joel Sebastian, Dex Card, John Records Landecker, Larry Lujack and little Tommy Edwards, Jim Stagg, Ron Britain, Barney Pip, Jim Runyon, Bernie Allen, Dick Williamson, Jerry G. Bishop, Bob Dearborn, John Rook, Ken Draper, and Robert E. Lee. These AM announcers, and directors of music and programming warranted the attention of huge audiences.

FM radio was in its infancy in the early 1970s, and the radio revenue stream from advertising reflected its newness. Had WYEN been an AM station in the 1960s, it may very well have had a tremendous following because of the quality of the announcers; still one can make a case for WYEN, Request Radio, 107FM. WYEN wasn't entitled to the *front and center stage* feel of WLS and WCFL based on its much shorter history. Yet, WYEN offered something the bigger-name stations skipped because they knew they were the major leagues. WYEN's vibrancy emanated from a new wave of mostly young broadcasters making WYEN their very first or second stop. They were unencumbered by the major league rules where only union workers touched the equipment. WYEN announcers had an eye on ratings but weren't consumed by ratings books. Mr. Walters pretty much knew his station fell with the thirty to thirty-fifth most listened to radio station in Chicago for every book. WYEN announcers found their voice and personality and completely ran their show themselves. They developed their style. Each announcer quickly came to appreciate a station where they'd showcase their ability in Chicago, the hottest market in the country. A number of these broadcasters you may recognize on your favorite radio station today. WYEN's proud family listed Jurek and Farella, Garry Meier, Greg Brown, Dave Alpert, Rob Reynolds, Bob Roberts, Wayne Allen, John Zur, Beth Kaye, Mike Tanner, Val Stouffer, Ray Smithers, Wally Gullick, Gil Peters, Bob Worthington, Mike Roberts,

Frank Gray, John Watkins, Roger Leyden, Paul Brian, Rob Sidney, Greg Stephens, Carmon Anthony, Kenn Harris, Louie Parrott, Jack Stockton, Adrian Sakowicz, Dan Diamond, Pat Lusk, Bob Walker, Mike Drake, Jayne Neches, Ray Baldy, Bob Andrews, Don Lucky, George Adams, Pervis Spann, Steve Kmetko, Jerry Mason, Bruce Davis, Bruce Buckley, Chris Devine, Bill Wilkinson, Mark Dixon, Bruce Elliot, Nick Kumas, Terry Flynn, and me. There were others, but this is a pretty good list.

What made *The WYEN Experience* a book worth writing is that it calls attention to a time when great, young talent worked without computers and cell phones. Despite a lack of today's tools, they excelled in every facet of broadcasting. The announcers identified with their listeners mainly because they were often from the same demographics. Their interest in WYEN's success stemmed from a sincere belief that they'd eventually springboard themselves into a more prominent position in Chicago and other major markets, and many of the broadcasters did.

I worked under Wilkinson's news direction until he left for a management job with Brown's Chicken. Bill's departure gave me a chance to answer directly to Walters.

Observations during my time at WYEN clearly showed heavy emphasis on music. The pressure fell squarely on the music director, and in the role, Reynolds seemed partial to The Carpenters and the Captain and Tennille. We manned our WYEN booth at Radio Days at Golf Mill Shopping Center in Niles, and we occasionally played softball against police and fire departments. Rick Saucedo was our station's favorite Elvis impersonator. America's "Ventura Highway" wasn't so much the song of choice in the mid-1970s as disco became mainstream, fueled by the "groovy" music of Donna Summer performing "McArthur Park," or The Bee Gees' "Night Fever" or Gloria Gaynor's "I Will Survive." Our listeners in the disco era spent their weekends line dancing in disco clubs, pretending to either be John Travolta in *Saturday Night Fever* or his dance partner, and they listened to the same music we were playing on WYEN. The music helped them "boogie" in their flower-print shirts, flared pants, and platform shoes, going to the most "gnarly" clubs for dance partners or for someone to play in backgammon. Quite a "heavy" scene with a steady beat driving "McArthur Park," "Turn the Beat Around," and "Le Freak," and the dance floor changed color, and the disco ball sparkled. No one wanted to "blow this taco stand."

The influence of music on society was never stronger than in the mid to late 1970s, but breaking news twenty-four hours a day was really in its infancy, ready to pop out for proud papa Ted Turner with CNN on June 1, 1980. Up until then, network television offered newscasts but nothing compared to what was to come with Cable News Network. WYEN provided perspective on national, international, state, local news and sports. This was for the most part what people listened to before the all-news radio and television networks changed our listening and viewing habits. Radio stations had taken their news seriously in the 1970s on both AM and FM, although after 1981 deregulation and the growth of CNN, the importance of FM news began dialing down and refocusing on the AM side.

I worked not only for WYEN as an anchor-reporter, but helped Chicago radio station WDHF/WMET news anchor-reporter Bob Roberts and News Director Dave Alpert. We had a reciprocal agreement in which WYEN news and WMET news covered stories and shared some sound bites. This way our news departments did not miss major events without a reporter present. During the years I spent in WYEN News, I attended some events that made national news, such as the news conference at Des Plaines police headquarters announcing the arrest of John Wayne Gacy, the first efforts of the Chicago City Council after the death of Mayor Richard J. Daley, the launching of Jim Thompson as the new governor of Illinois, the news conference by Skokie Mayor Albert Smith vowing to stop the neo-Nazis from marching in his town, the threat of the swine flu, the pardon of Tokyo Rose in Chicago, and Forest Park Police searching for the missing body of Michael Todd, the third husband of Elizabeth Taylor. I also took on feature assignments. I met and interviewed Shirley Temple Black at a Chicago hotel, played Ping-Pong and interviewed tennis great Bobby Riggs, did a feature story on up and coming comedians at Comedy Cottage, including the accordion-playing Judy Tenuta, discussed the 1972 Munich Olympics, swimming, and a new line of swimwear with US 1976 Olympic champion Mark Spitz, joked with TV star Tom Bosley of *Happy Days*, and talked Chicago Bears football with Hall of Fame running back Walter Payton at Woodfield Mall.

In the mid-1970s I reported on the assassination of four men in a Park Ridge office building, the JD Searle plant explosion and fire, and a plane crash near Palwaukee Airport (now called Chicago Executive Airport), all without the use of a cell phone. The handheld phone you can put in your pocket did not exist back then. Seventies radio stations used CB and shortwave if they could afford the equipment. Fortunately, in the 1970s we could still find phone booths scattered all over the place, typically at gasoline stations and in retail districts near restaurants. Everyone had land-line phones. A reporter need only use a little ingenuity and borrow a land-line phone in someone's business. I didn't have the moxie to borrow an office phone while covering breaking stories, so I looked for pay phones or rushed back to the radio station. Rushing back to the station wasn't really a good choice, though, because people liked the sound of immediacy and on-the-scene reporting. In chapter

8, Tech No Savvy Seventies, I tell how my introversion surfaced big time on the JD Searle Company fire in Skokie on June 27, 1977.

A single-engine plane crashed on a frontage road just outside Palwaukee Airport. As usual, Wilkinson called me at home, and I grabbed one of WYEN's very bulky and heavy cassette recorders, the Superscope by Marantz C-206LP cassette recorder, standard issue back in the day. It felt like a couple of bricks in a carrying bag, but you'd develop strong hands and arms over time, lugging the Marantz all over the place. When I got to the scene, or as close as allowed, I gathered information on the type of plane, condition of the pilot, and where the plane was headed before crashing short of its destination. I gathered a bit more information in an unorthodox but effective way, by listening closely to what other reporters were saying. I didn't have access to a phone on the frontage road where I could see a couple of businesses, but they all looked closed. Those were the moments I wished WYEN News had the equipment used by WBBM Newsradio 78 because their reporters had a bag's worth of technical equipment. WBBM's Fred Partido could be anywhere, and I knew this because he was often covering the same stories I responded to, except that I'd wander around looking for a phone, and he'd already have the report filed using his station's CB radio, and even though a few of his words might not air because of signal problems, he was still telling a story and doing it first. I gathered as much information as possible and drove back to the radio station and then reported the story with several versions and slightly changed leads. This gave the audio appearance of refreshing leads and finding new information. Today, I'd use my cell phone at the scene and do several updates, learning new information and reporting it minutes later. I'd start with reporting on details at the scene and then go on to anything new I'd learned, possibly the name of the victim or victims, and whether the plane was still there or had already been moved for investigative work by the Federal Aviation Administration. But more than thirty-six years ago, on-the-scene reporting was much harder on reporters without the WBBM-type equipment. We had standard issue alligator clips for clipping onto the metal tabs of the microphone inside the mouthpiece of a telephone handset. The clips looked similar to the clamps on jumper cables used for starting dead car batteries. These were used for sending back to the radio station sound bites taped on a portable recorder from an interview. Illinois Bell Telephone Company made most pay phones with mouthpieces that could not be unscrewed from the handset. This prevented someone from stealing parts inside the handset, but it also added to the difficulty of sending the interview or doing wraps over the phone. Wraps are audio stories with the reporter talking plus a sound bite.

Anything we saved from the on-scene interviews, we'd dub onto a reel of tape and use in promos for news. I can say now this manual method of saving tape was extremely time consuming, but who knew in the WYEN days that dubbing onto tape and then splicing and taping together was such a time waster. We'd put these reels of tape back in a box, and someone else would pick up the tape and do edits. You'd wind the tape around fast and could see the brown tape turn white from so many edits speeding by.

Wire services of the early 1980s experimented with computer program software. My newsroom, WAIT-AM 850 in 1981, used a pilot United Press International computer program of split screens with wire copy brought into the computer from around the world and viewed on my monitor screen. All that said, I did not slide easily into the computer mode because I'd never used one and was apprehensive. I still had deadlines every half hour to read my newscasts, and I didn't think the listeners would accept an excuse like, "I'm sorry, but this newscast will be very short because I can't figure out how to put my newscast together on this computer and print a copy." I hadn't yet thought of using the screen as a teleprompter. The station manager insisted on all the newscasts being typed directly onto the computer, so I went at it earlier than I'd normally start work the first post-typewriter morning at WAIT-AM. This was so reminiscent of the first time I tried to write news exclusively using a typewriter rather than writing stories on paper first. I never forgot this transition in my life because of the temptation of that pad of paper and pen on my desk just waiting for me to pick up my former tools. The professor hovered, waiting to land his voice right in my ears.

"Think on the typewriter," my Radio-TV professor boomed.

He was loud and direct. No confusion of message. If he'd been a piano teacher, and I'd been looking at the keys, he'd have slammed the piano cover on my hands.

He walked up and down the aisles in the classroom. The aisles were narrow. He'd see my copy and how little writing I had besides my name and the title of the story. Typewriters were on every desk. The clacking meant most of the students were thinking on the typewriter, or just faking, hoping he wouldn't go in for a closer look. I just felt like leaning over and putting my head on the typewriter, hiding my inability to translate my brain's thoughts to my fingers and the keys.

"Stewart, please don't write the story on a piece of paper first."

Suddenly, this wasn't the voice of my professor lightly admonishing failure.

For a moment of vivid clarity, I was a kid again with a new bicycle, and my dad was teaching me how to ride out front around our cul-de-sac.

Dad, I can't ride this bicycle, I thought.

"Dad, isn't this Lauren's bicycle?"

"This is your new bicycle, not your sister's."

I ran out of reasonable excuses. Did Dad think I was a mini Evel Knievel without the fancy jumpsuit?

"You'll stop me from falling?"

I did try to get his assurance because I was certain of one thing; I was going down. Failure was easy to expect.

The bike is too big, and I'm not reaching the pedals.

Dad was a product of The Greatest Generation. He stormed a beach on D-day amid large-caliber bullets raining down on Allied soldiers. Excuses were as foreign to him as the beach was in Normandy, France, that day on June 6, 1944.

I sat in my radio writing class staring at the typewriter, thinking back to my first bike ride and how Dad's resolve overwhelmed my trepidation. But I was angry at Dad. He was pushing me forward, and I wasn't ready. Yet I couldn't walk away, so he held the bicycle. He ran with me, pushing the bicycle only after I slowed down and the front wheel wobbled. Somewhere around the circle, he let go and stopped running. I looked behind me for a moment and then stared ahead and pedaled around the cul-de-sac on Osceola Avenue in Morton Grove, realizing I didn't exactly know how to get off … He got me a Band-Aid.

"You're a good soldier," I thought he said.

Thinking on a typewriter wasn't about skinning a knee on the street. Dad demanded I ride again and again, each time gaining more confidence. Dad taught me how to turn the unfamiliar into the familiar, and although he may not have known then, he prepared me for working on a typewriter and eventually a computer.

I began typing one thought at a time, interrupted only by hitting a wrong letter on the keyboard or trying to change my poor word choice into something that resembled the English language. The process seemed massively difficult, but a lot of people were just learning to use a computer. Sounds fairly juvenile, this excuse I had rolling through my head that I couldn't make a smooth transition from typewriters to computers, and this slowness could affect my work. What really were the choices? There were none. The only feedback I heard was the general manager asking if I got the hang of it.

These early computers and their pilot-type software programs began replacing IBM Selectric typewriters in newsrooms. The Selectric, in the WYEN newsroom of the 1970s, was state of the art. I suppose anything was better than my electric Corona in college with the sticking keys. I struggled in college every time I made a mistake striking the wrong key. I had to have a clean copy, and making additional copies using carbon paper behind the regular paper only led to carbon paper ink all over my hands. Don't worry if you've never experienced using typewriters and carbon paper. We're never going back to it. I won't even tell you about the fun you could have with milky white correction fluid. One drawback, though, in using carbon paper, correction fluid, or those correction strips—you'd just lose your train of thought working on your story.

Now imagine working in a small room with no windows though you do have a United Press International (UPI) teletype machine clacking away. You are seated beside a Selectric typewriter, a half sheet of yellow paper pinched under the roller. The AM/FM radio monitor on your desk is tuned to Newsradio WBBM 78, and a reel-to-reel tape recorder is connected to the phone and a speaker. Music from the John Zur show is barely audible in the newsroom with all the clacking and Bob Crawford's voice squeezing through the small radio speaker in one of his many WBBM reports from City Hall in Chicago.

So much of the focus in radio is on the voice and the ability to read well. That combination, when done well, will give you the best chance of success. Yet writing has a very clear place in developing someone into a triple threat (voice, reading, and writing).

In one of my college English classes, the professor strongly suggested each of us maintain a journal. He wasn't expecting entries that might mistake us for the great novelist Ernest Hemingway, but he did expect frequent writing, especially in times where we did not feel inspired. I wrote for a few months in my journal but began missing a day here and there, and eventually I switched to writing documentaries and news features and left the journal completely. All this writing eventually helped form hundreds of *slice of life* pieces called Stew's View. I wrote about everything that had an effect on my life or the lives of people I knew. Everyday things became important, and I fashioned those events into two-minute stories on-air on WAIT-AM 850, a Chicago area radio station that for a while was news and talk. Twenty years before Stew's View, I wrote about some of my WYEN experiences, and this helped me to remember things that I worked into *The WYEN Experience*. Not all my journal scribbling were in full sentences because I wrote quickly before I'd forget anything, and I never went back to smooth out the writing.

The Big Laugh

On February 10, 1977, I confronted a problem, and his name is WYEN announcer and part-time engineer Dan Diamond. The starter button on the top deck of the cart machine was malfunctioning. In my earlier newscasts, I would hit the button three times and be lucky to get the button to work. My 10 p.m. cast was interesting. Dan was fixing the button and started at 8:30 p.m. He kept saying, "Just a few more minutes," (five minutes to be exact). My last cast was partially engineered by Dan. I got to the story on former Cook County Corrections Director Winston Moore being charged with beating inmates. This was serious, but I had my moments with Dan in the room. He had taken a cart out of the cart machine and began putting another one in. This was all background stuff ... or was supposed to be, but Dan's arm and hand kept waving around. He'd been crouching on the floor below my vision, so seeing his arm and hand feeling around the cart machine looked so disjointed ... so funny, I did everything I could to hold back from laughing. I tried to keep focus on the news copy, but I kept looking at what he was doing. The rack with the three carts was unplugged. I cued him to play the cart he had helped me pop in. Nothing. Dead air. The cart didn't play because the power wasn't on. I could see his hands fumble for the cord and reach up, turning on a toggle switch on the cart machine.

A Good Day—Sundays in 1977

Sunday is usually a day of very little local news. But I have been lucky over the past few weeks. Stories that I can get actualities from pop up. Today I had three voicers, two from Nick Kumis, one from Bill Wilkinson, and three CTA El crash stories dealing with putting up a restraining barrier.

Garry Meier, weekday jock and Sunday announcer, told me that he is going on a two-week vacation in a week. Good for him. He keeps saying he's got to get to a bigger and higher-paying station. He's talented, and I wish him luck. Meier's journey will take him west to California.

I hope to get at least three years at WYEN. It's an excellent place to gain experience. After WYEN, hopefully a bigger station. I'd consider Florida or California, somewhere in a warm climate. Got to get away from the cold of Illinois winters.

Journal Notes

My notes mentioned Jerry Westerfield, Ron Leppig, Mike Roche, Wayne Allen, Rob Reynolds, John Zur, Ed and Carol Walters and their son Mike, and office staff Mary Ann O'Neal and Diane Finkler. I even wrote about myself during the period I worked at WYEN. Here's what I wrote:

Stew Cohen, news reporter, anchor, and news director, is mostly a serious person who wants to do the job correctly. Has periods of strong motivation and drive, but also has periods of not wanting to work hard. When he puts his mind to it, does as good a job as anyone on Chicago radio, but is suffering a swallowing problem right now. Doesn't drink or eat anything while working, says it affects his swallowing. Also is very wary of people who are sick with colds. Tends to get sick easily and loses his voice.

Ed Walters, general manager and owner of WYEN, is a friendly, buddy-type guy. He believes the equipment has lasted this long, it will last a bit longer.

Mike Walters, Ed's son, is a "Jack of all trades" or has tried to be at the radio station, handling all sorts of jobs. Mike started at the station right out of high school. He apparently has found his area in music programming. Also works with public affairs.

Rob Reynolds, music director and afternoon deejay, is a very friendly person who does a lot of outside work even as he goes about doing his jock shift. While Rob is clearly talented on-air and in the production studio, he is extremely messy, leaving records all over the place and scattering papers near the on-air console.

The brief comments on Reynolds, Walters, and the others were just a sample of the staff I described before I left WYEN. I also wrote about covering news conferences and the people I interviewed.

Right before my first newscast, I wrote the date, August 16, 1976.

"I'm putting together my first WYEN newscast. I do not know how long I'll stay in radio, but I honestly am not thinking long-term at WYEN because my dream is CBS News in New York. I'm twenty-two years old, out of college by a few months, working my first 'real' professional radio job thanks to Walters stopping at Golden Bear in Mount Prospect for breakfast."

Chapter Three
AN OWNER'S DREAM

You aren't a seasoned broadcaster unless you are fired three times. Never fired, you're an anomaly; once fired, you're a novice; twice fired, you're an apprentice; three times fired, you're a seasoned broadcaster and a member of the tortured family known as veteran broadcasters. As a member of this radio family, I can speak for our many fired members and say our family doesn't care how you lost your job; just the fact you had a radio job one day and don't have the job the next day is all that matters. If broadcasters were soldiers, we'd have medals pinned to our uniforms symbolizing the myriad ways we've lost our jobs. We'd also have patches for all the stations and cities we've worked in our career on the way to the market size of our dreams.

Mr. Walters hired and fired, believing fully his personnel moves were the best for WYEN. Not meaning any disrespect to Walters, he led a fair number of broadcasters toward the road to veteran status. He was also excellent at providing jobs, giving many people an opportunity to break into a field that held tight to the tenants of the movie, *Catch-22*. You stood before a radio owner, operator, or station manager and were asked about your experience, and you launched into how you worked several days a week at the college radio station and covered stories and ripped wire copy. You expected to "wow" your way onto the air, but the more you said, the more entrenched management became, and finally they said, "You don't have the necessary experience I'm looking for today." You couldn't gain the experience they suggested unless someone in the radio business stepped forward and offered you a position. Likely they wouldn't make an offer because you lacked professional experience. In the 1970s, catch-22 stopped many budding radio stars from ever starting.

Today, the "would be a radio announcer if only someone gave me a chance" is probably wealthy and retired, while some veteran broadcasters are overstressed workaholics and probably living paycheck to paycheck. They'll say they wouldn't want to do anything else. The affection that radio veterans at WYEN showed for their work wasn't lost on Walters. He looked for people with the same strong desire for radio that he had. His stable of broadcasters shared a powerful interest in entertaining and informing, a feeling that Walters knew very well. We put on the backburner our feelings about salary because we were convinced the experience was more valuable than the money, but over time, money's significance grew. We were gaining the experience and believed the money should reflect our growing experience level. Walters's WYEN came before automated radio found a home in many of the suburban radio stations of America. We were some years away from seeing a gradual homogenization of the medium as personnel inside the stations were eliminated for network-type programs, where the announcers were heard on dozens of radio stations. Also in the 1970s, radio was still a few years away from deregulation during President Ronald Reagan's administration.

My mother offered plenty of encouragement whether I played high school varsity tennis or read news at WYEN.

In the summer of 1976, a short time after graduation from SIU with a BS in Radio and TV, I assembled parts for phonograph needles at Shure Brothers in Evanston. This was where my mother, Shirley Cohen, worked in the purchasing department. She put a word in for me for a summer job on an assembly line. A couple days later, I sat in one of eight rows of tables with room for eight assembly line workers. I stared at tiny phonograph parts on a work table. Workers at each station had before them a small metal punch press bolted into the table. They held assembly tweezers and carefully picked up tiny metal cylinders crimped on one end. With precision and the watchful eye of assembly line foremen, workers set the cylinders inside their individual press in a specific way and then pushed down on a lever. Dozens of workers pressing levers, punching holes in cylinders, creating assembly line music never destined for the Top 40 on WLS. Foremen walked the room, examining a few finished cylinders here and there in the cups at each work station. I never filled a cup. I'm surprised Shure Brothers kept me for the two-month summer job.

My time was running out on the assembly line. Nothing looked promising—none of the radio openings in Iowa or Wisconsin where I applied. I used the *Broadcasting* and *Radio and Records* section in back of the trade magazines where news broadcaster jobs were advertised. Dozens of resumes and tapes around the country made no difference, though a couple of station managers in Puerto Rico and Alabama showed interest. A lucky break seemed the only real possibility, and this came at the most unlikely of places—Golden Bear Restaurant in Mount Prospect. While I worked at Shure, my brother-in-law managed the suburban restaurant. Walters went to his neighborhood Golden Bear before starting his day at WYEN.

Of all the Golden Bear restaurants, he came into my brother-in-law's.

Rick's Café from the classic movie *Casablanca* might have served similar food to the Golden Bear, but Humphrey Bogart's Rick may not have had a radio station owner as a regular customer. In 1976, my sister, Laurie, was married to the restaurant's manager, Terry Shindle. Terry and his staff developed a connection among their customers by super serving them. Terry walked through the restaurant with a hot pot of coffee for customers, filling Walters's cup. They talked about radio once Terry learned that Walters owned WYEN. Walters suggested Terry should have me send him a news aircheck tape and resume. This was my best shot at a job in radio and apparently my only real shot.

The light of broadcasting illuminated a path the moment my mother picked up her office phone and on the line was Terry, calling from the restaurant. He told my mother about his interesting customer. Now all my mother had to do was let me know, and she did so in the car heading home.

THE WYEN EXPERIENCE

She leaned over to the car radio and turned the volume way down. Something was up. She pulled out a piece of paper she had folded and stuffed in her purse and read her note. She told me the name of the owner, the radio station, and the phone number. Then she told me about how Terry talked to Walters.

"WYEN Radio, I'm Stewart Cohen," I whispered, looking through the passenger side mirror at the quickly shrinking Shure Brothers building.

My arsenal of seven-inch reels dwindled to less than a dozen, far fewer than the number of resumes I had printed. The tapes went out and never came back whether or not I asked to have the tapes returned. Stations usually dropped the tapes in a file drawer and then threw them all away every few months after filling a position. I put a WIDB aircheck reel in a box with my resume and cover letter and mailed the package to Walters at WYEN.

I made a serious mistake with my resume. Do not, I mean *do not* use a picture of yourself on your resume, unless you are applying for a job as an actor, runway model, swimsuit model, or model in general. Radio station managers have too many reasons to turn aside your resume. Now they're staring at your picture. What's a picture expected to do for radio? It's true today that your picture may end up on your website blog with your station's logo, but generally, a picture won't help your chances for a radio job. The station managers and program directors doing the hiring may hate someone who looks like you. They may think your clothes are outdated or see ominous shadows over your face, devilish redeye, or a slightly out-of-focus you. Keeping my photo off my resume wasn't something anyone advised. A photo jazzed up a meager amount of information. Skipping a photo and padding the resume may have worked better. Instead, I went ahead and had someone take my photo in a room with a pretty good light. However, my hair was way too long and looked uncombed. I needed a shave and probably a clothing makeover. I had on a leisure suit, circa 1976, rather than something more nondescript. My resume stood out in a negative way. Hey, it's radio! Who'd see me anyway? How about general managers, station managers, program directors, and news directors? Walters and other station owners had reason to chuckle and reason to turn aside the resume based on a less than professional photo.

A week went by, and I'd not heard anything. My excitement at the prospect of a call from WYEN dimmed by the day, and by the start of the second week, I had to know. During my lunch break at Shure Brothers, I headed to a room with a pay phone and started counting my change over and over. Counting wasn't so much a problem; rather I was extremely nervous, though no one else was in the room. I thought slightly delaying the call might ease my nerves, but the introvert inside me screamed.

Stop putting such importance to this phone call ... to this moment ... to the fact no other radio stations appear willing to hear from me, and make this call before someone walks in!

My pockets bulged slightly from enough change to keep Walters on the line for an hour, but I knew five minutes or less time on the phone was most likely. Besides, the way my heart pounded through my chest, I couldn't survive an hour phone call. Some crinkled paper in a garbage can on the floor near the phone was good enough for jotting down a couple of notes.

Hi, Mr. Walters, this is Stewart Cohen. My brother-in-law talked to you at Golden Bear Restaurant.

I wrote a few sentences on the note paper and practiced reading to myself. This wasn't going well. I wouldn't have hired myself—just too shaky! Still, I pressed on ... plopping a quarter into the phone, hearing the coin go through with its distinctive *click, click*, and then dialing the radio station's number, which was on a folded piece of paper my mother had given me.

"WYEN Request Radio," a pleasant-sounding woman answered.

"Mr. Walters, please."

"Who may I say is calling?"

"Stewart Cohen. I sent an audition tape a little over a week ago. I just want to make sure Mr. Walters received it."

"I'll check. Please wait."

I patrolled the phone booth, walking back and forth the distance the phone cord stretched.

"Mr. Walters did not receive a tape, or he can't find yours. Either way, I'll schedule you for an audition at the radio ..."

A bell sounded, signaling the end of the lunch break. I had to rush this along.

"Any day this week is good," I firmly told her, regaining some vocal strength.

"This Thursday works for Mr. Walters. Okay?"

"That's fine. What time?"

"Bring your resume by at 5 p.m."

"Thank you."

An audition? Oh, no! Well, I've got a couple of days before then.

I got the directions to the O'Hare Office Plaza and drove there in my orange Vega. My resume sat in an envelope on the front passenger seat. Two security guards patrolling the parking lot drove up to my car. They seemed fairly animated rolling down their windows, sticking their heads out, and although I couldn't quite understand what they were asking me, I answered, "WYEN." They pointed at the entrance.

Don't goof this up, I repeated to myself every couple of steps from my car to the entrance and down the stairs to WYEN.

THE WYEN EXPERIENCE

The voice at the front desk matched the woman I'd talked to on the phone. She introduced herself as Diane Finkler. She was as pleasant sounding in person as she was over the phone. Diane looked to be in her twenties, stylish and fit, and presented exactly the right appearance for a station playing WYEN's brand of music. The listeners were thousands of *Finklers*.

"Stewart," Diane said without hesitation, "this is Mr. Walters."

Mr. Walters put out his hand a few feet away as he walked toward me and we shook hands.

"May I have your resume?" Mr. Walters asked. "I'll take a look and I'll be back in a second to give you a reel of tape for the audition." As he left the room, I looked around.

From a small front office to a little larger office area, the room opened into the largest room at WYEN, just outside the main studio.

A very tall and thin guy wearing a cap ripped copy from the wire service machine and set the stories on a table where I'd been leaning.

"Hi," I said, waiting for him to acknowledge.

"I'm John Zur," he answered, smiling. "Are you here for an audition?"

The depth of his voice amazed me. You don't expect such power from anyone tall and thin. I knew right then, *This is professional radio*. Another man with such vocal power was SIU Professor Richard Dick Hildreth. He was professional from his perfect enunciation to his command of the language. Dick's voice amazed me too as I heard it for the first time in a radio class he taught in 1975. Hildreth's deep-rooted resonance imbued hope in his classes that he could teach us to speak with the same vocal qualities. Maybe Hildreth had Zur as a student?

"Yes, I'm here to audition."

Zur was the WYEN evening announcer—a perfect fit, I thought—but he sure made me feel as though my voice hadn't matured.

While I waited for Walters to return, WYEN people moved around outside the main studio. I noticed something unusual. Most of the staff members were white. Walters returned with a reel of tape.

"You've got mostly white guys here," I said.

Walters stared at me for a moment. Oh, no, I had blown my only real chance. Really, I didn't have an opinion on the station's on-air staff. I'd learn later that Walters had a solid office staff of women. On-air though, I hadn't met the lone female broadcaster. Walters calmly told me he would hire the best talent he could find … and that's it. I escaped a close one!

A SHOP OWNER HOPES ZUR'S LEGION OF FANS VISIT HIS STORE.

Walters had one of the announcers grab copy and gave the stories to me. I practiced and then told him I was ready. The practice time was critical. Practice for fifteen minutes, and Walters might think I had trouble reading; practice for ten minutes, and Walters might think I had to look up a few words; practice for five minutes, and Walters might think I wasn't serious about preparation. I guessed at seven and a half minutes of practice.

"I'll have you read into a recorder in the production studio, right there." Walters pointed at the room straight ahead.

Seemed odd I wasn't especially nervous. This big moment had so much riding on it. Maybe the half-dozen WIDB radio auditions under Strom helped. I scooped up the copy and walked into the production room and got some instructions on recording my audition tape. A little over seven minutes later… "I'm Stewart Cohen, WYEN Metro News." Except … I didn't say *Stewart*. For the first time, I used Stew. Leaving behind all the things associated with "Stewart" felt right. I hoped a new chapter had begun, and a new name or actually a shortened version seemed appropriate.

I turned off the microphone, hit stop and rewind on the recorder, grabbed the tape and copy, and found Walters waiting nearby. I never heard the newscast played back. What mattered was what Walters thought, not whether

I thought the cast was good or bad, though if someone gave me the audition tape later, I'd have taken all the tape off the reel, stepped on the tape, and thrown the whole thing in the garbage. Those early newscasts have always been just too hard to stomach. But Walters must have had a strong stomach for all the audition tapes he'd heard over the years, and one more he put in the pocket of his sport coat. I should call back in a week was his advice. Walters did as he promised. He listened to the recording and must have felt it passable. The evening I called the radio station was one week to the day I left WYEN thinking I'd never see the station again.

In two and a half years at WYEN, I earned the equivalent of a master's degree in communications. No one passed along a diploma, and sadly I realized I stayed too long. For most of my time, I worked forty hours a week and eight hours on Saturday and covered at least two stories a week on the road. But I hadn't fully guarded against burnout because I didn't know its symptoms.

One afternoon late in 1978, Walters asked me, "How many hours do you work a day?" He must have sensed I had cut back slightly. I would leave an hour earlier during the week. I didn't know how I'd explain this to Walters. My job was in jeopardy. Yet I couldn't lie when he asked for my hours, and I remember he said, "Okay," and seemed satisfied. Walters left the newsroom, and I continued gathering news, not thinking anything more until a couple of weeks later. He asked me to his office. I could see he looked pained. I stood just inside the doorway. Walters looked at me. The whole purpose of this meeting was clearly expressed on his face. I knew it. No point in forcing him to do something he abhorred. He just wasn't comfortable letting people go. Walters was much better at having a conversation with a coworker about radio, sharing the enjoyment of broadcasting. That's a subject he could talk about for hours, but this necessity of the general manager job of firing people never sat well with him. I thought maybe I should say something to save my job, but I didn't say a word. I knew the motivation to continue at a high level just wasn't inside me now.

"Are you letting me go?" I asked, slightly relieved to finally move this forward.

"I appreciate the work you've done in the news department, and I'll help you in every way I can."

Walters was genuine. I had no reason to harbor any anger toward him. He gave me a chance to do what I had trained for in college, and I was grateful. Still, my legs were slightly wobbly, probably because I wasn't yet a veteran of broadcast firings. I told Walters I'd been thinking about enrolling in a cooking school. But I wasn't completely quitting on radio.

"Could I make tapes and get a reference letter from you?"

"No problem," he answered. "I'll tell you what, Stew, read the news next week. I generally don't do this, but I can't think of a reason why we can't go a little longer, until I find a replacement."

He was giving me time to put tapes together.

"Thank you, Mr. Walters."

We both felt slightly better.

Most of the week was uneventful, though I counted down my days at WYEN by producing airchecks. The best of the newscasts were recorded to a master reel with features and news stories. I boxed my best work with a resume, targeting the boxes for stations with openings in news found in *Radio and Records* and *Broadcasting*. With two days left at WYEN, I caught something in my throat and could tell my voice was going. By the final day on Friday, I had just a rasp of a voice and couldn't do news or say good-bye to anyone.

More than fifteen years after I quietly left the WYEN Des Plaines studios, I saw Walters again; this time he walked into my newsroom in Crystal Lake. Talk about shock. Had I been chewing gum, I'd have swallowed; had I been drinking, I'd have bit into the glass; had I been trimming my mustache, I'd have cut my lip. I still wonder whether he knew I was working there. Station owner Jim Hooker introduced us. He didn't know our history. Walters was unmistakable in his overcoat, looking hardly older than at our last meeting in his office. So many things I'd practiced saying to him if we'd ever meet again. Most of it had to do with convincing him the person he fired did not even remotely resemble the person he was seeing now. Here was my chance to say all the things bottled inside for the longest time, but Hooker moved Walters through the newsroom surprisingly fast. I didn't know I'd only have a few moments.

"Hi, Mr. Walters."

"Hi, Stew."

"It's been a long time," I said.

I realized this wasn't what I had in mind for the conversation of two people who've not seen each other in years, but sometimes what you've played through your head doesn't always spew forth from your mouth.

"Yes," he agreed.

"Good to see you again," I warmly responded.

I felt no animosity toward Walters back then, and I certainly did not now.

"Stew, you're working at a good radio station," he said.

"Thanks, I know."

But how did he know this was a good station, and what was he doing here anyway? Was he looking to buy another radio station? Just a few weeks earlier, legendary broadcaster Howard Miller walked into my newsroom, just as Walters was doing. Hooker was waiting. I could see he had an agenda for Walters, and a long newsroom stopover wasn't part of it.

"Stew, I've got to go. Good seeing you."

"Good seeing you too, Mr. Walters."

Could I have said less? I didn't take my eyes off of Walters until he shut the door behind him. I had hoped he'd walk back into the newsroom, remove his overcoat, sit in a chair, and listen to all the things I've done since WYEN. Why did I feel I needed his approval? I couldn't shake the feeling of letting him down by not fulfilling the trust he had placed in me at WYEN. Even years later, I felt a need to show him his trust was beneficial in my continuing in radio.

Jump ahead fifteen years to November 20, 2010.

I last saw Walters's widow, Carol, in February 1979, but now I'm working on *The WYEN Experience*, and former WYEN Program Director Ray Smithers is visiting Carol, and I'm interviewing him for this book. Carol held a little welcoming party for Ray.

With Mr. Walters, I'd talk about radio, never about personal matters. He was best and happiest this way, it seemed to me. But I found a willing listener in Carol, and I knew she was a direct pipeline to Ed. By the second week of work at WYEN, late August 1976, my radio confidence quotient (if there really is such a thing) was so low I did not want to have a conversation with Mr. Walters asking me why I was not cutting it on-air. Seeing him disappointed was more than I could handle without my confidence quotient falling even more. I made mistakes reading news, and I had this swallowing problem where I couldn't read a sentence without forced swallowing, and it just did not sound smooth stopping and swallowing and then reading some more and swallowing all over again. I went to Carol. She was in her office at WYEN. I felt awkward seeking out the boss's wife, but I felt her sensitivity level was higher than her husband's. I told her I was trying but was disappointed in myself and had been fighting low confidence, which showed in a swallowing problem. I promised to do better on-air than she and Ed heard so far. But I think I did not look convincing, not sure in my own mind that I could find the answer. Yet her insight brought me back.

"Ed hired you because he believes you will do a good job," she justified and confided. "Don't worry, Stew."

"Thank you, Mrs. Walters."

Mrs. Walters's perfect answer didn't melt away my problems. I struggled for several weeks, but she bought me more time to overcome. I never forgot Mrs. Walters's sensitivity.

No one knew Walters's dream better than his wife. The couple shared their struggles with the business of radio, shared the good times, and shared the feeling of accomplishment in giving young broadcasters a start. I asked **Carol Walters** to take *The WYEN Experience* back some sixty-two-years. She told her story to her grandson, Dan, and in turn, he turned Carol's words into a story of the love they shared for each other and for radio.

Long before there was ever an Ed Walters in Chicago radio, there was Edward Piszczek in a North Side diner. I met him there in 1950. I was having lunch with my sister, Ruth, and our girlfriends. Ed and his cousin walked in, and after briefly consulting with each other, they approached us. Later, I'd find out what they were discussing. Ed wanted to know about me. His cousin said I was his classmate Ruth's little sister, and then Ed declared right there that I was the girl he was going to marry.

We both grew up on the same block of Armitage Avenue in the Bucktown neighborhood, one of the many Polish enclaves on the city's North Side. It's fashionable now, but it was miserable then. Blocks and blocks of simple, blue-collar Polish families seemed perfectly content to do their labor every day, go to church every week, and never make it out of the old neighborhood. That wasn't me. I was a budding artist with cosmopolitan interests. I saw myself doing big things. Yet I lived in the basement apartment of a crumbling three-flat with four siblings, an overworked mother, and an alcoholic father. I thought it below me and couldn't bear the thought of living that kind of life all over again with a family of my own.

Ed made his case to me, but I wasn't interested. Plenty of neighborhood boys had asked me out, but I turned them all down.

Sure, they'd plead, and Ed certainly did, but I told him as I told the others before him, "I won't date boys from Bucktown; I won't date Polish boys!" Having several in my family already, "I won't date anyone named Edward."

I thought living by these rules would be my ticket out of miserable Little Poland. How fortunate I would be to break all three of them on the same man! My friend Harriet wanted to date his cousin Frank, and she insisted that the four of us go out in a group. We rode in a car together, but I just stared out the window until I felt I should try to start a conversation.

"What is it that you intend to do?"

"Well, right now, I'm going to a technical school to be a refrigeration man."

"Humph. If you were going out with me, you'd have to be going to school to be a lawyer, a doctor, or a judge." I was unimpressed.

Growing up poor, I wanted nothing more than to rise above my station. I needed to know that I would never again have to live among extended family in a shoddy and cramped three-flat, or worse yet, be evicted from one. He continued.

THE WYEN EXPERIENCE

"Don't laugh at me, but it's not what I really want to do. I'd really like to be in radio."

"Radio!"

Here I thought I was talking to another neighborhood kid with the same low aspirations as the other boys I wouldn't date. But this was someone who dreamed big, just like me.

"Well! If you really want to be in the radio business, then why tell me you are going to plumbing school?"

Before I knew it, we were putting him through Columbia College together. The radio business took us around the country. For his first job, Ed was a salesman. He sold radio public service announcements on the road to the small-town stations that needed them to keep their licenses. We must've gone through every little town from Maine to South Carolina, saying our Hail Mary prayers as we rattled up and down the Appalachians in an old Plymouth convertible. Finally, Ed found steady work in Watertown, but I found rural Wisconsin just as unappealing as Bucktown. We moved to Ottawa, Illinois, where Ed hosted a morning show, and then back to Evanston, where he worked in sales again. Ed's big break came at 100.3 in Chicago, the old WFMF, where he quickly rose from salesman to host to general manager. But by 1960, Ed decided that it was time to stop running a station on behalf of Marshall Field's and start running one of his own, like he had long dreamed of doing.

Of course, buying an existing station was out of the question, but there was only one open frequency left in Chicagoland: 106.7 FM, licensed not to Chicago, but to the northwest suburb of Des Plaines. FM radio was not the dominant force it is today; its creator jumped out a window to his death because he thought his life's work was a failure. But Ed was a visionary and believed in the power of FM while the world around him clung to the old AM band. To get that frequency, it took us years of fighting with rival prospective owners, Washington lawyers, and FCC bureaucrats, and when WFMF's owners found out their general manager was applying for a license of his own, they fired Ed on the spot. He brokered radio time independently while I groomed dogs to make ends meet until we were finally awarded the last license. We went into business with one of Ed's advertising colleagues, Jerry Westerfield, who financed the purchase with the help of generous loans that no bank would ever give today.

We got to work right away, building our tower on high Arlington Heights farmland and building our offices and staff in Des Plaines, our legal city of license. Ed's first move was to hire a program director, and he chose Ray Smithers. Ray was a familiar face whom he had hired at WFMF and now gladly poached from them. He knew that Chicago was a crowded market, and to set us apart from the established downtown stations, Ray devised an unprecedented format: all-request radio, where anyone could call in at any time to request almost any song. While press for the new "Request Radio" circulated, Ed and Ray ran a tight "mock station" to prepare us for going on the air. It sure was hard work for an audience of zero, but ultimately we would sound just as polished as the other stations, as if we too had been around forever. When it came time to flip the switch, we had more listeners than we ever could've imagined.

It still wasn't easy. In the beginning, Ed and I did everything to keep the station running. We did sales, bookkeeping, and even reception, all in the name of keeping our costs down. Dealing with the FCC and what we felt were their rigid, out-of-touch regulations often made our lives a nightmare. And we always feared that no matter how much we were billing, it wouldn't even be enough to pay our electric bill. But WYEN would be a joy in so many respects. Ed wanted with all his heart to entertain our listeners, and we did. And we were just as entertained by our station as our loyal listeners were. Along the way, we met so many celebrities who passed through our station, not to mention the many colorful and unforgettable personalities who came to the office every day. We may've been in a suburban office building, but we were in radio, and radio is, after all, show business.

COMEDIAN PHYLLIS DILLER AND MR. AND MRS. WALTERS ENJOYING THEIR EVENING TOGETHER.

The show must go on, so I asked Mrs. Walters's daughter, **Jackie Walters Heinlein**, to give *The WYEN Experience* insight into her relationship with her father.

As a child, I adored my father. When he would wake up early to put in his long days at a station in downtown Chicago, I would wake up even earlier to spend time with him before he left. I looked forward to weekends, when we would run errands together. He quit smoking at my behest because I wanted him to live forever. I was—and still am—proud of the sacrifices we made as a family in getting WYEN off the ground. This is not to say my faith went untested.

Though I always believed in him, sometimes the rough patches made me vocalize my concerns. We couldn't move during the process of applying for a license, and so we spent years in a Prospect Heights house where the flooding of the Des Plaines River system would ravage our home year after year. From the time my dad began lobbying for his own radio station to well after he finally got it, our lives were full of so much crushing stress that I, as a young girl who had no lifelong love for radio, finally had to ask why we were in the game, why we made these sacrifices, why we had to stay in a house that flooded. On a day when my parents were discussing the difficulties they'd had at the station that day, which was not yet on the air, but still in "dress rehearsal" mode, I broke down and finally asked my father what makes him so different? How is our station going to be different from anybody else? Dad could have taken the question as mere venting from a frustrated daughter, but he answered me calmly and with the candor of a young idealist who doesn't speak a single word without meaning it.

"Our station," he said, "is going to be part of this community. I don't want to just play songs. We're going to be part of this community, and the community is going to be part of what we do. Nobody else is doing that here, and we will. And that's what will make this work." My dad truly believed in his dream.

I had dreams of my own, however. My lifelong dream was to work in special education, devoting my life to the lives of those affected by disabilities. I would go on to teach in a special education program in Woodstock and Crystal Lake, and the gratitude of the children and parents I worked with made me feel like I'd chosen a worthwhile and noble job. When my dad asked me to leave my job and help him at WYEN, I had a difficult choice. Eventually, we struck an amicable agreement. I would leave Crystal Lake for WYEN, but I would work in the station's public affairs department, which to that point was only a nominal post.

FCC regulations call for all radio stations to air a minimum amount of programming deemed relevant to the "public interest" in or around the city of license, and the FCC does not believe that playing popular records is in that interest. You'll traditionally hear these shows on Sunday mornings. Most stations treated their requirements like a necessary evil and produced only the most minimal and perfunctory programs for the public good. Under my watch, however, public affairs grew into an integral department of the station. I thought back to my dad's earnest desire to run a true community radio station and put myself into that department just like my dad put everything into the station. To me, public service was about much more than filling legal requirements. I revered the unique power we'd been given and used it by providing beneficial information to our listeners.

I began, as most idealistic newcomers do, by cleaning up the last person's mess. Inside the glass walls between reception and the studio was the old conference room, which had been carelessly repurposed into a storage room. Not only was it going unused, it made a poor impression on guests, who would gaze through the glass to see the deejays, but find their views obstructed by overstuffed boxes and crates. I took an entire weekend to clear out the junk and redecorate the room into one celebrating our station's famous guests and successful events. The conference room would become the guest room where our radio personalities interviewed local politicians and civic figures for our weekly show, Community Insight.

CHICAGO MAYOR JAYNE BYRNE WITH JACKIE AND HER MOTHER, CAROL IN CHICAGO.

My greatest achievements in public affairs were the March of Dimes walks and our participation in the local branch of the annual Muscular Dystrophy telethon, with our radio personalities acting as television hosts. This wasn't just typical fundraising drudgery. One year, we hosted an event featuring a local sandwich shop that erected "the world's largest sub sandwich log cabin" for charity!

As expected, our listeners were intrigued and made large donations to say they could eat the walls of a house. And to this day, I have a special place in my heart for Ricky Tummillo, the son of one of our listeners. He was born with a rare skin disease, and his mom, a big fan of the radio station, came to us for help. Together, with the help of local celebrities, we had our very first radiothon, and it was a huge success.

Leaving behind my dream of special education was hard, and I'll always remember the people whose lives I touched. But I have no regrets. I met my husband Kenn at WYEN, and together we have two beautiful children. My daughter, Elizabeth, is studying to be a pediatric nurse, sharing my passion for helping others. My son, Daniel, is a brilliant writer and speaker, just like his father and grandfather. He likely would have been my father's ideal protégé, sharing all his passion for every area of the radio business.

I miss my dad every day, and I never stop thinking of how dedicated he was to living his dream in radio. Many of us never get to live the lives we always wanted to lead, and he was a rarity in this respect. He not only spent his life doing what he wanted to do, but had many successes and achievements along the way. His life in radio was an amazing one. I'm fortunate to have been part of it.

Harold Piszczek had a close relationship with his older brother. Long before radio became Ed's passion, Harold remembers how his brother was always there for him.

"I know from personal experience my older brother was always very supportive and nurturing. I was ten years younger, but he always included me in the things that he did when he was single and living at home with the family. When his buddies were around, they were working on their cars, getting them ready to run on the weekends, so the garage was a good meeting place, and I was always invited."

Growing up, Ed never indicated an interest in the field of radio.

"I would know because Ed always let his intentions be known to his family. He was very easy to talk to and hold a discussion. If I had a problem, whether I was down about grades, we'd get together and talk, and I always felt better. When I was deciding to start my own engineering business, I talked to Ed a lot as he proved a great sounding board. Though we were thousands of miles away, he always remained a good brother, and we stayed close until he passed away."

Harold knew love and respect seemed to dominate the lives of the Piszczek family, not just between brothers, but stretched comfortably to the whole family, reaching the heart and soul of niece, **Diane Finkler**. *The WYEN Experience* had Diane writing that one year after she graduated from high school, "Walt" hired her.

He knew I didn't know what I wanted to do with my life, so he gave me a job as a receptionist at the radio station. I quickly nicknamed him Walt because it was hard to call him Mr. Walters. He was my uncle.

Walt stuck. Everyone in the business started calling him by that name. Soon Walt and I knew that we were a great working team. He advanced me to programming. This was putting commercials, public service announcements, and anything that needed to be scheduled for the day's broadcasting shows. I loved it! I did it for about eight years. Then I was advanced to secretary to the president for my final six years. Yes, we were together for fourteen years. We could actually finish each other's sentences and thoughts.

Walt was like the dad I never had. He always looked out for all of his employees, but especially his family. There were several of us working at WYEN.

In our early years, we were poor. We depended on all the press parties, free lunches and dinners. Looking back, those were probably our best years. I learned so much about the broadcast business and business in general from Walt. This was like going to broadcast school for me. I have carried all he taught me into all of my working endeavors. As office manager of a pediatric office, there isn't a day that goes by that I don't think about Walt. Something always will turn up, and I have to draw from experiences at WYEN.

We always had fun but were always professional when need be. Yet I was the practical joker in the office, and at times, my little practical jokes worked as a great stress reliever. Ed had many stressful moments with the Federal Communications Commission.

Walt's strong points really focused in on bringing talented people into WYEN, among them Garry Meier.

As I write this, I shed a tear for a man I loved dearly, my Uncle Ed. I miss him so.

Thank you for everything, Walt. See you on the flip side.

While Diane was family, **Ray Smithers** felt like part of Carol and Ed's family. Smithers told *The WYEN Experience*, "Ed was one of the nicest people I ever knew in radio."

Even years later after WYEN, as creative director *production guy* at KMPC in Los Angeles, Ray recounted, "I would help Ed out with programming ideas from time to time."

Walters hired Ray as his first program director because no one knew Request Radio better than the creator of the format. Smithers and Walters were friends since they were together at WFMF, owned by Century. (Ed and Ray began working at WYEN several years before WFMF changed call letters to FM-100.)

THE WYEN EXPERIENCE

A FRAMED PICTURE OF SINGER JIM CROCE WATCHES OVER DIANE FINKLER'S SHOULDER AS SHE AND NEWS DIRECTOR BILL WILKINSON PREPARE TO GREET VISITORS TO THE WYEN STUDIOS.

In the next chapter, Ray tells *The WYEN Experience* how he got WYEN to compete with the more established stations in Chicago. But in the early days of WYEN, Ray faced a problem announcer, and his name was Ed Walters. Ed was the second person on-the-air after the December 1971 kick off of 107FM. Smithers says Gil Peters signed on first at midnight so he and Ed could make sure the station's transmitter "wouldn't catch fire" before Ed started the Morning Show. But quickly Smithers recognized that Ed was much better as general manager with a hand in sales than on an announcer shift.

"Ed did the Morning Show, and he was a disaster. Ed wanted so badly to get back on-air. He was a deejay as a kid, and his blood was full of radio, and he wanted to be on-the-air again."

On December 7, 1971, Ed sat down at the FM board in the WYEN studio and aired his 6 a.m. to 9 a.m. Morning Show. Ray knew Ed wouldn't do too many more.

> ON-AIR
>
> *The music of Ferrante and Teicher this morning on the Ed Walters Show on WYEN Des Plaines, Stereo 107FM. This is a new*

morning for you. We're kind of new here also at this dial setting, only I believe what is the third day now? I'm getting confused myself when we look at the days we've been trying. We've been here trying for about nine months but finally gave birth to our little baby here. WYEN Stereo107FM, hopefully we'll be here every day of the week, twenty-four hours a day to entertain you; keep you informed on everything that's happening throughout the Chicago Metro Area.

The time now is twenty-nine minutes before the hour of seven.

"Ed's show got worse and worse. I was doing all the promos, of course, and I'm taking airchecks of the show. I got nothing to put promos together, and I'm listening, and I'm saying, 'Oh my God, this is a disaster; it's not going to happen.' I really had to twist his arm to get him to give it up."

Ed stayed on for a week on the morning show, finally ceding to Ray's call to find an announcer he believed was a better fit for the prime WYEN air slot.

Ray took a moment to describe Walters as a person.

"He was just one of the crew at the station and never separated himself in an ivory tower. We all ate, drank, and slept radio, and when three people or more share passion, it is a sort of mutual love."

Ray once answered a question years later about Ed and simply said, "You know, I never in all the years I knew him ever once heard him raise his voice."

Ray writes that Ed was just a regular guy.

The bathrooms in the office complex where the station was located were down a hall and shared by other tenants. On this one particular day, very early in the station's life, only Carol, the announcer, and I were in the station. Ed was there too, but he'd been down the hall using the restroom. When Carol answered the station phone and screened the call, she discovered the guy on the other end of the line was a commissioner with the Federal Communications Commission. She left the station's suite, ran down the hall, opened the men's room door, and shouted, "Ed, the FCC is on the phone—come quick!" As Ed came dashing back to his office, I happened to be sitting there when he picked up the phone, out of breath, and said, "Sorry, I was taking a crap!" He turned a bit red, and the conversation continued without ever returning to the comment. Ed knew the most important thing for a broadcast owner that since has been forgotten. As program director, I had lots of off-the-wall ideas, and my creative inventions seemed to know no end.

I would sit down with Ed and pitch him with the idea, and he nearly always told me that if you believe in it, try it.

Ray observed when radio went south; he saw countless managers hire creative people and then put them in a box and not let them do the very thing they were hired to do.

"I always thought Ed cut such a wide swath for creative people simply because he was not threatened by them, because he also was a creative soul himself."

Frank Gray followed in Ed's footsteps.

"Ed gave me the morning show, so I think he thought enough of me to do that. I thought Ed was a really hard worker. He was overstretched, handling the program director position after Ray left and all his other duties. Ed must have been a good sales manager back in the days before WYEN, and he was sharp enough to get that last FM signal, but so stretched in doing everything. I think there were shortcomings with Ed and money because of what he had to pay to get the station, and shortcomings from a marketing perspective."

"Got to crack that nut," **Rob Reynolds** remembered Ed's trademark line, and his trademark movement of scratching his back on a door frame.

The WYEN Experience had Rob reflecting on his relationship as music director.

"We would talk for hours about radio. I felt he was very wise in terms of street smarts. He was down to earth, extremely tenacious, would pursue relentlessly what he wanted. He fought hard to win the last radio license."

However, Rob believed Ed was an incredible procrastinator. Rob said one evening Ed came into the station while he was getting ready to go home for the evening.

"Ed told me he had to fly to Washington DC for his radio license renewal. He was going to sit down and fill out his application, and I'm now helping him, and he says we're 25 percent news, and I said there are five to six minutes of news an hour, not even close to 25 percent. I stayed with him until 4 a.m. I thought for a guy fighting that hard to get the last license, he did not appear to me to work hard at keeping the license."

Reynolds's point on whether Walters worked hard or not at keeping the FCC license is certainly open for debate. However, what's very clear is his consistent effort at maintaining friendships. One such lifelong friendship was with **Bill Jurek**.

Bill visited Chicago radio stations in the 1960s while he was a high school student. Among the stations, Jurek and his cousin worked their way through the Carbide and Carbon Building on North Michigan Avenue and dropped in on Walters at WFMF 100.3FM.

"Walters used to talk to me about applying for an FM frequency. He went through angst to try and get the station on-the-air."

Walters had contended with strong opposition for ten years—a history Bill found helpful in discussions with Ed over his hopes to also start a radio station someday.

"I ended up working for him at WYEN, and he was always a big supporter. Walters and WYEN were very responsible in giving young people a chance."

Plucked right out of college, **Wally Gullick** was a testament to Jurek's observation.

"I'll never forget Mr. Walters for giving me a shot after I finished college. By hiring me, I could get a job in Chicagoland, and yes, fifty thousand watts was a kick for me. I wasn't close with Mr. Walters socially, but there was mutual respect from the start, and I left on good terms. I'm indebted to Ed Walters."

Jack Elliott, a popular broadcaster today in Oklahoma, used his birth name of **Ray Baldy** at WYEN.

"I didn't spend much time talking to Walters, but he was always really nice. He was always smiling when I saw him. Everybody always talked about how Walters was very frugal. As a young guy getting into business, I didn't care. If I got a check for $28, I was happy. I was just happy to be in the business. This was something I wanted to do all my life."

Walters's generosity may not have been in the form of salaries for his talented staff, but Ed was abundantly generous in other ways. Weekend announcer **Philip Raymond** didn't need coaxing to tell the story in *The WYEN Experience* of what transpired one cold Saturday night in Chicago.

"The clutch cable on my 1981 Plymouth Horizon snapped, and I had no way to get to my WYEN shift. I lived in an apartment in Chicago's Albany Park neighborhood. Walters picked me up in his Cadillac Fleetwood and drove me to WYEN in time for my shift. He later drove me downtown to the Hancock Center for my day job as a technician at Channel 66. Back then, Channel 66 was doing pay TV movies under the name Spectrum. I did not know Ed well at all, yet he was willing to help me out when I needed it. For what little I knew of him, he seemed to be nicer than most radio station general managers. He may not have paid well, but he did what he could. Walters was a good guy."

A good guy, a nice guy, and a personable guy are descriptions of Walters. **Wayne Allen** served as program director in the mid-1970s and recalled Ed as a "very secretive guy."

"I remember when I was interviewed by Ed. He wanted me to start on weekends. I was excited about that. It was just a thrilling time. However, Ed seemed to hold a fair number of closed-door meetings."

Walters's favorite commercial voice on WYEN gave the station an incredibly rich sound on spots. **Jack Stockton** was incomparable.

"Ed was a nice, personable guy, but he was caught by the finances of owning a radio station. He had a fairly powerful radio station, but on the fringes. When he built the station, he thought he'd get the premier accounts, but he didn't get those accounts because the numbers didn't get up there. You didn't earn much, but poor Ed wasn't able to pay those big bucks. He had to rely on the main-street merchants. He was in a hard place, and we would talk about that and how the station could progress. Ed's dream was to have WYEN as a real competitive force in the Chicago market, but this was a suburban station and an FM during a time FM stations weren't all big as they are today. Ed was disappointed with the industry and his dream. Maybe if he had a million bucks in a suitcase, he might have pulled it off, but that wasn't in the cards."

Another big voice for Walters came from radio sales at WYEN. **Sherri Berger's** career took off with her myriad voice work, but early on, WYEN was her first sales job. She handled traffic and other areas in radio before moving to the Chicago area.

"Ed was sort of fatherly to me. He was very nice and sweet and encouraging."

Sherri knew he took a chance on someone with no sales experience, "but he never shied away from telling me how proud he was of me."

Walters's pride for **Bob Worthington** increased as Bob's career shot straight up from national recognition on Solid Gold Saturday Night.

"As I remember Ed Walters—Ed was *a pacer*! You could always find Ed pacing back and forth in the lobby, in the office area thinking of what he had to do next! He did kind of remind me of Mr. Carlson on *WKRP*. Ed was a good guy. I was very impressed on how hard he worked and his background as former president and general manager at FM 100 in the 1960s before taking on ownership at WYEN."

Car references might work well for **Paul Brian** and his pit stop at WYEN. Today, as the spokesman for the Chicago Automotive Trade Association and Drive Chicago on WLS 890AM, Brian recognized Walters as sales driven.

"Ed taught me the lessons of being cooperative with the sales department. He encouraged me to find ways to do things rather than find ways not to do things." This is a philosophy Brian lives by to this day. He felt Ed was a good match with Sales Manager Ron Leppig.

Kathleen Cahill is sales manager of Chicago area radio stations WZSR-

FM and WWYW-FM. She started in sales at WYEN and eventually landed a position as general manager of WLIT-FM.

"Walt talked a lot about his past as a general manager at a downtown radio station (FM 100). He seemed a bit disenchanted of the powers in place there. Owning and operating WYEN was Walt's way of proving he could go off and run a radio station. He was pretty charged up about that. Ed was a nice guy."

No one was more *charged up* than **Tony Salvaro** calling the harness races at Maywood Park and calling in race results nightly on WYEN. Though he rarely visited WYEN, Salvaro knew Walters fairly well.

"He'd drop everything to spend time with me. We liked each other's company."

Salvaro described Walters as a close guy, "because when you got him laughing, he'd touch your shoulder, and that was good. I had a ton of respect for Walters."

Salvaro was rarely seen at the radio station, but someone just as invisible to staff was Jerry Westerfield, the other half of Walt-West Enterprises. His mother helped provide start-up money for WYEN, but very little was known by the staff about this copywriter partner of Walters.

Westerfield would go right into his office next to Walters's office and stay for a good portion of the day, and he'd leave without mingling with the staff. Most of the time, you wouldn't know he was in or out because he kept his door closed.

Westerfield called me into his office once. Someone came into the newsroom and said he asked to see me. I didn't want him waiting too long, so I followed into his office. I noticed right off, conditions were the opposite from what I'd seen in Walters's office. Westerfield had few papers anywhere, and everything seemed very organized.

I stood in front of his wooden desk. He had a camera, a pen, and some other items on his desk that in its entirety told nothing about his personality or character. He made pretty good eye contact and seemed pleasant enough, though I was slowly learning his displeasure over my story of a hot-air balloon accident in which a man plunged to his death in one of those balloons that tangled with high-tension wires. Must have been my delivery of the story where he had trouble. I may not have given the story the compassion he felt it deserved. Instead, my interpretation made the story sound just like any other story. He felt I must in the future recognize life lost and give the appropriate treatment. I appreciated the fact that someone was actually listening closely and thanked him for his comments. That was pretty much it. He'd made his point, and we were done.

The final time I saw Mr. Westerfield, he had insisted the on-air staff pose for individual promotional pictures. He got out that 35-millimeter camera on his desk and took pictures of the entire on-air staff for some publicity shots we could give out at remote events where the staff would occasionally go and broadcast. He took those pictures of Meier, Allen, and a few others near the pond in the middle of the office complex. We did Radio Days at Golf Mill in Niles and brought our stack of headshots. Westerfield printed a couple hundred pictures for each of us. I wasn't one of the announcers whose picture was in high demand. By the time I left WYEN, I still had almost every picture Westerfield gave me, and I still have the pictures today. I found Allen's promo picture in my batch of pictures. Don't know what happened there.

Walters and **Smithers** were in total agreement over the authority of the program director position. Ray had complete control and total trust from Walters, which was important for Ray in developing WYEN into Request Radio.

"Anyway, Jerry comes to me one day, and he starts pulling albums, and he's got a stack of them against his stomach, and he tells me to play them."

Smithers liked Westerfield, so he took the albums.

But wait a minute, Smithers thought further, putting the albums back in Westerfield's hands.

"Go ask Ed about this."

Smithers knew Walters would support him because of their deal, but also because this was All Request Radio.

Ever talk about someone behind their back and discover they've been listening the whole time? In such an embarrassing moment, you feel like crawling into a hole and waiting until the person leaves. **Reynolds** joined friends at Hackney's in Wheeling, and they got to talking about WYEN. Rob started telling them things that had bothered him about the station. When the group was finished with their meal, they prepared to leave. Rob noticed to his chagrin that Westerfield was sitting in the next booth and could have heard everything he said.

What were the odds? Rob thought.

Reynolds wasn't sure what Westerfield heard, but he never said anything, so maybe Jerry did not hear a thing or maybe he had enough class to let it slide. Rob knows he was lucky that night.

Program Director **Jerry Mason** shared something fairly obvious with Westerfield. So obvious, it mostly could have gone unsaid, but Mason remembers Westerfield was quite a character. "The first time I met Westerfield, he says, 'We have something in common … we're both named Jerry!'"

Chapter Four
BIRTH OF REQUEST RADIO

OFF-AIR:
"Hi, this is WYEN Request Radio. Do you have a request?"

"I'd like to hear 'Time of the Season.'"

"Ah, great song, The Zombies' 'Time of the Season.' What's your name and where do you live?"

"My name is Sue, and I live in Wheeling."

"Sue, your song is just ten minutes away. Have a great day."

ON-AIR:
"A request from Sue in Wheeling for The Zombies' 'Time of the Season.' This is WYEN Request Radio."

The role Ray Smithers played at WYEN was critically important to the station's success in the 1970s as Request Radio. Smithers gained experience at suburban and Chicago radio stations with myriad formats and positions within the stations, but mostly as program director. Walters sought this brilliant programmer to shape his dream into reality.

My interview with Ray and the chronicling of his career provide a window for peeking at the building blocks of WYEN. Ray would go on and offer radio listeners good music with a distinctive sound and presence.

Lester Vihon bought WNWC, Wonderful North West Communities, from Bob Atcher, the long-time mayor of Schaumburg and a country Western singer. This Arlington Heights-based radio station, 92.7FM, had the benefit of location. The station was in the center of a rapidly growing area in Chicagoland. Vihon recognized the benefits and sold WFMQ. Smithers worked for Vihon on WFMQ and WKFM.

"A little coffee pot of a station," was how Smithers remembered WNWC. "Vihon and his GM Wayne Smith decided to build a new building and leave the old house located on Rand Road. Just as construction began on the new building in a nearby industrial park, I got a call from Ed."

Walters had listened to Ray and liked what he did on-the-air. They met for the first time at a Red Balloon restaurant.

"Ed offered me a job at WFMF, the number-one FM beautiful music station in the market."

In 1947, WFMF stood for Marshal Field. By the time Smithers and Walters met, Howard Grafman headed the Century Broadcasting station. Grafman would be a significant player in the life of both Ed and Ray.

While at WFMF, Ray got a call from Wayne Smith, who was the general manager at WEXI. He said Walter Mack had purchased WNWC and it would become WEXI. Ray asked if he could meet with the new owner. Mack was a local Cadillac dealer in Mount Prospect. Smithers lived on the sixtieth floor of Marina City at the time and invited Smith and Mack to his apartment where he had put together some production samples of what he thought WEXI might sound like. After a few days, they offered him the program director job, and later he became station manager.

"Mack was a brilliant business man, and today I rely on what he taught me in the daily operation of the two companies I own with a partner."

An employee of WEXI and Ray began production music at a recording studio on Michigan Avenue called Stereo Sonic Studios. With the backing of Bruce McGuinn, his friend's father, Ray and Bruce bought the studio and renamed it RayMac Studios. During that time, Grafman decided he no longer wanted Walters as general manager at WFMF. Grafman knew Walters tried to talk to the Federal Communications Commission about securing the last available signal in Chicagoland. "So Ed was given the boot."

Chief Engineer Ivan Bukovsky, Ed, and Ray became very close friends at WFMF. Ray felt this friendship would be the perfect combination for starting a new radio station. He knew Walters's struggle securing the final radio license in Chicagoland dragged on with him no longer working as WFMF general manager, but instead landing a job as sales manager at RayMac Studios.

"During that period, I was working overnight at WIND. It was very strange to broadcast out of the same studio where I listened to the man that made me want to be on radio, Howard Miller."

Also strange, Ray realized, was when Westinghouse moved the station's studios from the Wrigley Building where it had always been.

"I had a very emotional moment doing the last broadcast *ever* from the Wrigley Building. A chill went down my back as I handed the 6 a.m. hour to Robert W. Morgan."

Ray will never forget saying for the final time ever in that studio, "This is WIND Chicago."

Smithers recalls Florida was starting to look better and better.

"Once my partner's backer decided that a constant monthly loss was not a good thing for his financial future, it was decided to sell the business." Smithers had a brother living in Fort Lauderdale, so off he went to the land or oranges and sunshine.

"I did not do well with finding work in Fort Lauderdale, though I didn't try very hard, and decided to return to Chicago. About two months after I got back in town, I found out that Ed had been granted his station."

Ed called Ray with an important question.

"We're going on-air, and we're building the station now. Do you want to be a part of it?"

Ed knew the answer, but he felt good asking because this moment was very long in coming. Ed hired Ray two months before he turned on the transmitter.

What would be the WYEN format? Smithers was all about creating a big splash for the new radio station while he and Walters recognized one wrong decision on the format could waste all the money spent on securing the license. That's why Walters hired Smithers even before the transmitter was installed. These were serious discussions plotting out how they'd establish a format and gain listener loyalty. Smithers knew WYEN could not just switch formats every few weeks if Walters did not see the numbers right away.

"We have to make a big noise with a little station," Smithers maintained. "We're joining eighty-eight frequencies, and how do you make a splash in a pond that big?"

The brilliant programmer had two ideas he felt could separate WYEN from all the other stations in the Chicago area.

"Every single cut (song) would be something someone asked for, along with one to two lines of text. You know, I want to play this for Mel who is not feeling well."

This idea was the basis of All Request Radio developed by Smithers before

THE WYEN EXPERIENCE

WYEN signed on. The other idea needed some arm twisting before Walters would buy into it.

"We'll go to all the major ad agencies in Chicago, and we'll say we're putting this new station on-the-air. We'd tell them WYEN has a lot of power and will be very important."

Very few people knew what Ed and Ray planned.

"We planned to run free spots for our clients, but we only wanted Class A accounts, such as Pepsi Cola, General Motors, and Coca Cola, and we had a contract they'd sign. We gave them twelve weeks, and the contract included the fact they could not reveal they were not paying for it."

Smithers knew this would give WYEN the best quality "right out of the box. We came out, and it sounded like we were already covered in gold, and it worked beautifully."

From the Walters Heinlein collection.

WYEN went on-air sounding rich with quality spots, but the station only had a couple of local accounts that Ed talked into coming on the station, and then the staff had to work to produce those equally as good as the national spots.

"The stir created in the radio business was nearly immediate," Ray acknowledged. "Major stations began calling the agencies asking how come Coca Cola is buying this thing that just went on-air and doesn't even have any numbers showing its share of the audience."

Amused by the secrecy, Smithers recalled, "The reps would say, 'Nobody knows anything about WYEN, and you as an ad agency are buying them?' The agencies sworn to secrecy to keep the deal intact for free air time could only say one thing to the question ... 'Yep.' None of the agencies ever squealed. It was really amazing."

Finding talent for on-air announcer day parts and news did not prove impossible because word got around very quickly. As program director, Smithers began looking, knowing he'd have some time because Walters planned to have the announcers in place and trained before the transmitter turned on.

"You don't run an ad in the *Chicago Tribune* saying deejays wanted, but you know what the network was like back in those days. It took very little for the grapevine to go into action. I'd call someone that used to work for me and mention an audition and see if there's interest. We put a staff together, and oddly enough it stayed together a very long time."

WYEN had Zur and Peters, and for news John Watkins, who went on to great things. Walters and Smithers also had the son of the late WGN Flying Officer Leonard Baldy. Ray Baldy worked in the WYEN traffic department. Filling the slots with talent turned out fine, except for the part that had Smithers redirecting Walters's on-air efforts into a more behind the scenes owner-operator role. One of the biggest challenges seemed to become an ongoing concern over the equipment used by the announcers. Smithers remembered very inexpensive equipment.

"We had home turntables, and everything was kind of funky."

Equipment aside, long before WYEN, Ed and Ray created a deep and lasting friendship.

"Ed made a lasting impression on me long before the WYEN years that deepened our bond—two radio guys with big dreams."

A few months before the air date for WYEN, Ed invited Ray to join the station, and for Ray, it was a no-brainer partially because Ivan was already there as chief engineer, and this would bring the *gruesome threesome* back together again.

Ray believed WYEN made it because the station had a cast of hopefuls.

"Nearly every night, staff gathered at the station or at a local watering hole. Everyone liked everyone, and we all had only one (at the moment) goal: to get this station on the map. Ed and Carol would often be around for those staff hangouts."

The truth was as Smithers saw it through his staff, "Those that are reaching for greatness do not watch a clock or ask for overtime; as a matter of fact, they simply don't want to go home."

In Ray's conversation with me at Carol's home, he leaned forward to tell me he's kept a secret until now. For *The WYEN Experience*, Ray wanted to come clean on Margo, the voluptuous secretary answering request calls in the afternoon during his show from 3 p.m. to 6 p.m. Listeners used to call her during Ray's show and try just about anything to get her to send them a photo. Some would flirt with her so much Margo had to cut off the call.

"Margo was actually the boss's wife, Carol."

This is something Ray could not tell listeners. "Call now, and one of the owners will answer the phone because we are so poor!"

Ray admits Carol was a great sport about the whole thing, and often to this day, he still calls her Margo, and she still giggles.

Ray received a call one day from the great Art Roberts who had taken over WGLD and asked to meet for lunch. Lunch ended with him offering Ray middays at WGLD Solid Gold and operations manager at a salary figure that he had not seen since doing overnight at WIND. Ray felt he had to say yes to Roberts but also hated to leave the team he had built at WYEN.

"Today I know that an amazing number of those people went on to great careers and great lives. I will never forget any of them and still am in touch with a few, thanks to Facebook."

Ray is proud to have remained a close friend of Mrs. Walters.

"Ed is someone that helped shape who I am, and many others, so he goes on and in the most important way, in the hearts of others. Ed, if you are listening, thank you for letting me be me."

A Smithers's stager and the impressive jingles defined the station's character. The stager gave WYEN a dramatic and distinctive sound with just a hint of mystery.

A winter surf on Lake Michigan.

The city skyline by night.

It's all part of the good life in Chicagoland.

And so is the great music of WYEN.

On September 23, 1972, Billboard released a story: "WYEN Goes All Request." The article quoted Ray on the all-request phone format. He told Billboard that WYEN was an MOR Station (Middle of the Road) with a slogan, "The Station You Can Talk To." The article listed the original lineup of announcers.

In the wonderland of WYEN, down the stairs, through the door, we'll soon visit with such characters as egghead Frank, Mike the Roche, and G.I. Bruce. We first come face-to-face with a raspy Gil. Any similarity to Lewis Carroll's characters in *Alice in Wonderland* is purely accidental.

CHAPTER FIVE

INAUGURAL AND ORIGINAL BROADCASTERS

Gil Peters

On December 3, 1971, "De Natura Sonoris No. 2" premiered at the Julliard School of Music. Composed by Krzysztof Penderecki, the piece translated to "On the Nature of Sound." Almost eight and a half years after its premier in New York, Penderecki's music chilled millions of people watching the classic horror movie, *The Shining*.

December 3, 1971 also brought to our ears for the first time music from a completely new source, WYEN Request Radio. Instead of haunting music of the Penderecki compositions, the music of WYEN brought "Joy to the World" and reveled in "Sweet City Woman" and let it be known that "It Don't Come Easy."

Peters drank tea for his raspy throat, hours before his scheduled *station kick-off* shift, but he knew he should call in sick. He'd caught a nasty case of laryngitis. Gil sounded terrible, and he knew it. Yet he understood the significance of the moment. All the practicing the announcers logged behind the controls had to mean something to Gil, so he had no intention of missing his shift. He cleared his throat; otherwise, he'd kept quiet for the hours leading

THE WYEN EXPERIENCE

up to midnight, giving himself the best chance for delivering his audience-pleasing voice that Smithers and Walters knew would go over big on WYEN's overnights. Peters waited for Smithers's taped sign on, Westerfield's message to the new listeners, and UPI News. Gil loaded the single cart machine.

"I pulled forward on a lever, bringing the capstan to the ready point, before hitting the play button."

Gil made sure the volume was up on the pots. He pressed the cart button. Uplifting orchestral music with trumpets, French horns, and trombones slid in and out with the hint of snare drums. Up and under Smithers's voice, the volume lowered on the driving force of the music.

The dream envisioned by Ed now played out with Ray's voice.

"This time, radio station WYEN begins another broadcast day."

The dream filled the air through the studio monitors and transmitter, radiated out of the towers, and reached thousands of radios in homes in Arlington Heights, Des Plaines, and beyond. This great production of a radio station sign on scored at midnight, relieving the stress for all those behind the planning of this moment. A full orchestra of sound heralded its beginning and trumpeted the fact that Request Radio was here to stay. With a voice registering between a tenor and bass, Smithers too heralded, *WYEN operates on the federally assigned frequency of 106.7 megahertz with an affective radiated power of fifty thousand watts vertically and fifty thousand watts horizontally. The WYEN studios are located in the O'Hare Lake Office Plaza at 2400 East Devon Avenue in Des Plaines, Illinois, with tower and transmitter located at 11 West Dundee Road in Arlington Heights, Illinois, and is owned and operated by Walt West Enterprises Incorporated.*

The music Smithers chose for kicking off WYEN could easily have celebrated the nation's birthday on the Fourth of July. Reaching a crescendo, the pounding drums, crashing symbols, trumpets, flutes, and tubas stopped flat. Peters waited three seconds. He separated each piece by only a moment of silence. Then Gil played the message by Westerfield.

"On this broadcast today, December 3, 1971," starkly contrasted the start of WYEN's broadcast day. Jerry's voice did not receive musical accompaniment. His was not typical broadcast quality, sounding akin to accidentally turning the bass way down on your car radio while turning the treble way up and adding extreme nasality to the mix.

"On behalf of the management and staff, I would like to take this opportunity to thank the many people who have cooperated to make this new radio station a reality. A special note of thanks to the late Birdie Westerfield."

Perfect timing! UPI's distinctive sound of clacking teletype started on Westerfield's final word.

"From the World Desk of United Press International, this is Tom Wendell."

Peters cued a record on one of the two turntables, loaded another cart, and had the cassette deck ready too.

"President Nixon is now in Key Biscayne, Florida," Wendell continued in his UPI report. "He took along his top economic advisors. They will be putting the final touches on the new 1972–73 fiscal budget during a weekend review."

Peters let Wendell get within ten seconds of finishing, and then he flicked the key above his microphone pot, cleared his throat one last time, and made the mike pot hot.

"From the World Desk of United Press International, this is Tom Wendell."

Peters turned down the UPI audio pot as the clacking teletype ended in tone.

"This is WYEN Metro News." Gil threw himself into his delivery though his voice had suffered. Emphasizing each word, he compensated for his strained vocal chords.

"Lieutenant Governor Paul Simon and his two opponents for the democratic nomination for governor met with democratic slate makers this afternoon in Chicago. Simon drew three rounds of applause. Chicago weather—tonight fair and a low in the twenties in the city and upper teens in the outlying areas. Right now, O'Hare has twenty-three officially, Midway has twenty-eight degrees."

Peters hit another cart, potting down his microphone volume momentarily.

Smithers set the mood for Gil's overnight show.

"Ray had in mind a Franklin McCormick type of overnight show with piano playing in the background. He wanted me as a laid back host being much older than I really was."

Ray told us in his hypnotic voice amid soothing music of violins and flutes, "The city is dark and asleep."

So quiet was the night, crickets were the loudest sound surrounding Smithers's voice, which rose above the insects and accompanied the sexiness of the night music.

"The crickets sing their song of night, and you have a rendezvous with the all-night music of WYEN."

The music bed's distinctive television soap opera quality reestablished itself, fed by violins layering mystery on top of longing.

Everyone crowded into the room awaiting Gil's first words. They prayed for a long relationship with the listeners. The Walters and Westerfield families hoped their request radio station had a place in Chicago radio. So much rode on Gil and other announcers following with a seamless quality product. Gil didn't appear nervous. He was solid and versatile, good at announcing music and weather, joking with his audience, or reading a few minutes of news every hour, though he felt the show was really set up for an older host.

"I worried about the cassette deck going off speed, sometimes playing the music so fast while I was talking that it sounded more like jack-in-the-box music going at an insane speed."

Gil noted the time, 12:05. The heart of his show began with a bed of piano music playing from a cassette player.

From the Walters Heinlein collection.

"This is Gil Peters on the all-night program on WYEN. I'd like to welcome you to our all-night show and welcome you to WYEN. Why not give us a call if there's something you'd like to hear tonight, being this is our first night on the air. The telephone number here is 297-8430, and the area code is 312 if you're calling from far away—and there's a good possibility of that. Once again, the area code is 312, and the number is 297-8430."

The first song, an instrumental, played on the turntable. Peters listened through his headphones, satisfied the music sounded the way Ray intended.

Peters recalled, "The music overnight was supposed to be some dreamy, loungy thing, maybe Henry Mancini, but no one knew exactly what it was, but it wasn't Montovani."

He cued another record on the other turntable. He'd picked an album from the top of a stack on a table near the microphone. Gil had pulled the first hour's music though he made room for requests on the very first overnight. He used to start out the show with the song "Songman" by Cashman & West, 1972.

"Songman, sing your song until the morning comes. I can listen as long as you can play ..." Peters smiled, remembering the words defining the overnight guy. He loved Bonnie Koloc and still does, just as his musical idol is John Denver. Peters ended his show with Denver's "60 Second Song for a Bank." He even parted his hair down the middle as Denver did so many years ago.

For *The WYEN Experience*, Peters relates his own story, beginning with his exposure to WYEN as a student of Columbia College.

I was a radio fanatic and was visiting my school friend, Frank Gray, at WEXI in Arlington Heights. He said he had something to show me, took me to the back door of the station, and threw the door open and said, "There it is!" The "it" was the tower for WYEN. This was a new station, and better still, a station I assumed was hiring. Frank and I both wanted jobs at a bigger, better station, and we figured that WYEN was the answer. As we were to learn, the answer is really only good if you know the right question.

When WYEN was ready to hire people, Walters brought in Smithers as program director. He was a bit of a local celebrity in that Ray is the one who started WEXI, which for a suburban station with little power became a real force in FM radio. I believe that many of the later music-heavy contemporary radio stations came about because of Ray and WEXI. I got an interview with Smithers, who intimidated the hell out of me, though I know he wasn't trying. I wanted to be back on the radio so bad I could taste it.

My experience was at another small FM, known as WKAK-FM in Kankakee. That station is now known as 9FM. Its dial position was 99.9. There, I learned how to do all of the important jobs at a radio station—you know, taking out the trash, washing the studio windows to get the tobacco smoke off, putting big reels of automated music on the machine, and also doing some broadcasting.

I don't know if my experience got me in, or if Smithers thought I had talent, or if I came at the right price (minimum wage). I was hired to do the overnight show. Most of the original group of WYEN announcers spent a couple of weeks there awaiting FCC approval to go on-the-air. And while we were there, we did typing, record filing, just about anything we could. We were going to be great.

And then the call came. We would go on-the-air officially at midnight

THE WYEN EXPERIENCE

December 3, 1971, and I would be the first official person on-the-air. I say official because we had some music and some sound effects of waves lapping the shore (I think) and Ray announcing that something brand-new was coming. But in the wait for the station to sign on, something unexpected happened.

I went on a date a few days before the station signed on. That wasn't the unexpected thing. What was, however, was that we went to a drive-in movie. *Butch Cassidy and the Sundance Kid* was playing. I drove my Corvair to the show but didn't keep the heat on because of the Ralph Nader scares of people dying in the cars. And it was cold. By the time December 3 came around, I had laryngitis and sounded horrible, but I could still talk.

That night of December 3 was really exciting. Walters was there with his wife, Carol, and Westerfield hung around for a while. They had all of their friends listening. The friends started calling and asking us to play songs for them, and thus the idea of Request Radio was born. I think they just liked the idea of people calling in. But it worked. And the phones rang off the hook.

A lot of unusual things happened with Request Radio. As the only person who worked in the station overnight, I had to answer the phones, write down the requests, find the records (yes, we played records ... those big black plastic CDs) and do the news, read commercials, everything. But as a result of answering the phones, you got to know your listeners. Most of them were terrific people who had certain favorite songs you could depend on playing for them night after night.

I made friends with Barry from Lincolnwood, Angie from Chicago, Marsha from Glenview, Theresa from Rolling Meadows who worked at Dunkin' Donuts, Cheryl from Niles, and Sheldon from Evanston. I also met Tom, the Chicago police officer who drank too much, and Joe, the Des Plaines police officer, and more people than I can remember talking to but just can't recall their names. I had dinner at some of their houses and had a wonderful time with them. I still have a table in one of my rooms that Cheryl gave me when she was moving.

Marsha loved John Gary and took me to a concert of his. Angie was a big Dave Major and the Minors fan and treated me to a dinner show they did. Theresa brought me donuts and gave me a dog. It was truly awesome.

Perhaps the best call I ever got, though, was from Cathy in Morton Grove. She wanted to hear a song entitled "Come Down in Time" by Elton John. I played that but then did something that I used to do on occasion. I played the same song by Lani Hall. I liked the contrast of two artists doing the same song. I also liked Cathy. We've been married for thirty-five years.

Then I wound up—as did many of our jocks—playing a role in an illicit affair. We would get requests from "The Cookie Man" and a woman who had another nickname. We didn't realize they were requesting for each other. They were married, but to other people. This was an affair that apparently grew steamier as time went on. We learned what we were participating in when "The Cookie Man's" wife called the station in tears.

Another time, I was answering the phones when a woman called me to say she was going to commit suicide. I kept her on the phone for a couple of hours talking to her, and she decided not to do it. About two years later, I got a call from the woman's sister to tell me that she had died of an illness, and that she had written a note that I should be called to thank me for spending the time with her. Humbling!

Along with the good times, of course, there was also the frustration of working for a small radio station. Money was one of the biggest challenges. I started at minimum wage, which was $1.75 an hour. Even by 1971 standards, that was dirt cheap. Making it worse was that the management decided to pay us in pay periods, which gave us checks on the fifteenth and thirtieth of every month. That is when they would pay us for two weeks of work. It didn't take a genius to realize that we were only being paid for forty-eight weeks a year. One of my best financial days at WYEN was when the minimum wage increased to two dollars an hour.

In 1975, we were all broke, but the station was doing well. The program log had a lot more commercials on it, so we knew there was money coming in. We thought about becoming a union shop but didn't really know how to go about it. So we decided to be our own union and demand more money from the boss. I was still on overnights and worked at a print shop to make enough money to pay the bills. The meeting was in the middle of my sleep time, so groggy and angry, I joined my fellow announcers by going in to see the boss and demanding our due.

We walked in, and Ed probably sensing what was ahead said, "I assume you are all here for a reason." I figured someone would say something quickly because we were so insistent on meeting Walters. His challenge met with silence; no one was saying anything. Not being one to ever keep my opinions to myself, I started talking. This was a big mistake, but I didn't know it at the time. At the end of the meeting, Walters said he would look at the books and have an answer for us next week.

Walters's answer was a series of raises over ninety days. I went from $72.00 a week (he only paid me for the six hours a night on the air, no prep time, just six days a week for thirty-six hours) to a grand total of $117.50. Not great, but a real boost, though not for long. The other shoe dropped when Cathy and I got married and went on our honeymoon. I came back to work on a Saturday in mid August. Walters was waiting for me. Because of my role in the salary meeting, and because I couldn't get in to work in an unusual April blizzard, I was no longer needed. As it turns out, it was not unusual for people to be let go while they were on vacation. Don Lucky was even fired while he was in the hospital.

My tenure at WYEN ended in August of 1975.

There's always a lot of complaining among radio people. I think it's the "artist" mentality that we carry with us, even if we're really not that artistic. Perhaps it's ego. But while we are disgusted with management, we still take with us a lot of lessons.

I learned how to do a lot of things at WYEN that improved my work as a broadcaster, such as improving my skills as a board operator, developing a personality for broadcasting, and taking advantage of a funny moment. Nothing gives you experience in adlibbing as when a poster comes un-taped from above the clock and flops down over it as you're back timing to a network newscast, or when the cleaning lady comes in while you're on the air and starts banging the garbage cans around, or when the cleaning lady hits a switch with her feather duster and dusts you off the air. It is those teachable moments that helped me to grow as a broadcaster, and which I will never forget.

On a personal side, I've maintained contact with several former WYEN announcers. Gray is still my best friend. Frank works with me part time as my IT guy. I am hooked up with Dave Alpert on LinkedIn. I work at WGN Radio 720 with Garry Meier, although during the blizzard of 2011, we actually worked together for the first time in his time there. I've run into Paul Brian a few times and worked with him at WCLR (now WTMX). I've worked with Greg Brown who spent a little bit of time at WYEN. I've talked to Steve Kmetko a few times in the past few years. He was really big on E channel on cable. At WYEN, I believe he did traffic and other stuff for a few weeks, probably as an intern.

I have been a radio survivor. Since WYEN, I have worked at WZBN-WKZN in Zion as a news director, where I learned tons about pulling stories out of my behind. I helped to move WCLR from an automated format to a more present music station. I was at WCFL during one of the worst formats ever constructed, Lifestyle Radio, reported for the Mutual Broadcasting System, NBC News on radio while at WKQX and WMAQ. I was news director at WJJD and did fill in at, among others, WJMK, WCKG, FM-100, WZSR-FM Star105.5 (as Christopher Michael for Stew Cohen), WKRS, WXLC, and a true highlight of my career, working as the fill-in editor for Paul Harvey. Currently, I am the president of Sound Targeting, Inc., a radio syndication/public relations company in Morton Grove, Illinois. I am host and producer of two daily syndicated features—The 60 Second Checkup and Rural Health Today. We also syndicate A Senior Moment with Clark Weber. I am a voice on a nationally syndicated half-hour show Viewpoints, which is produced by MediaTracks Communications in Des Plaines, and work as a part time news anchor at WGN. I marked forty-five years in radio in 2012 and plan to keep on talking until I can't do it anymore. I also pray for the ability to recognize when that is.

Frank Gray

Peel the shell off this hard-boiled egg and find out how a guy goes from spinning records in the early days of his career to working Internet Tech for the American Egg Board. Long before Gray transitioned into IT, he followed Walters, the morning show host, after only a few days as the regular morning show host waking up listeners in Chicago's suburbs from December 1971 to 1975.

"I was one of the original crew, hired at first to do public affairs programming on WYEN and a jock shift on weekends. Then Walters decided he didn't want to do mornings and own the station and be there twenty-four hours a day. That's when Ed and Ray asked me to do mornings, which I could not turn down. I was in my last semester at Columbia College, and I remember rushing out of the Des Plaines station at 9 a.m. after my shift, yet always being an hour late for class in downtown Chicago."

Frank found out about WYEN while he worked at WEXI.

"All of a sudden, this new tower starts going up in the station's backyard. I asked Mike Drake to tell me about the tower."

Not only was Frank working with his friends, Peters and Drake, but he'd have a chance to work with Smithers, whom he admired for programming WEXI.

"I can remember when Ray brought in the jingle package designed to be a spin on the old WCFL jingle. The jingles were lush and descriptive. We even had little vignettes or what we called the intros or introduction to the shows back then."

Gray says he became WYEN music director by default.

"When we first signed on, we were playing albums on turntables, but these were old albums of Percy Faith and John Gary and whatever strings, so I started to bring in contemporary music at the time, and I took the jingles and edited them down from a minute to shotgun jingles so they would be played between songs."

During his WYEN years, Frank served in the army reserves. When he finished his service and came back to work at WYEN, he wasn't doing morning radio anymore because someone else was filling in. Paul Brian had stepped in for a while but left, and Frank returned to mornings. Eventually Gray too left WYEN for *Smoking Oldies* in the windy city.

"I went from doing my own show on WYEN to doing news on WFYR-FM, an automated Chicago radio station. From a market perspective, I advanced, but I was no longer doing my own show."

Gray stayed for fifteen years at WFYR, and in 1995, he became the director of IT at the American Egg Board.

"Clark Weber and several other great broadcasters motivated me to go into radio because there were professional communicators and personalities then. I was a child of 1960s WLS and WCFL. But today, it's not the same business. There's very little personality in the business today."

Bruce Elliott

Looking at one of the sexiest women in the world and hearing what she had to say about dancing, acting, and singing was by design. The program director didn't even have to try convincing Elliott to sit down with Joey Heatherton at the Playboy Club for a WYEN interview. She had entertained troops for Bob Hope's USO Shows during the Vietnam War, appeared on television shows with Dean Martin and Perry Como, and made the rounds of Playboy Clubs. Many of Bruce's interviews for the radio originated at the Playboy Club and at Hugh Hefner's Mansion and Grotto.

"I was a little intimidated by the mansion itself. The people didn't intimidate me because I was too stupid to be intimidated, but here was a cool place—down to the grotto and the whole routine. I even saw Hefner halfway across a room."

Smithers found in Elliott a guy whose ego prevented him from feeling fear, even among very successful entertainers.

"I had a chance to go to these places like the Hyatt Regency O'Hare and see great acts and interview them. One of my favorites was Jerry Butler. He was doing great stuff. Butler had been working with a young composers' workshop to see if he could help develop their musical skills. That was a cool thing because he had years of huge successes from songs like 'Let It Be Me' and 'For Your Precious Love.'"

Elliott built solid relationships with entertainers and the people behind the venues. For nearly three years, he offered insightful radio interviews with such stars as Lou Rawls and Bill Withers. In the year "Lean on Me" charted for Withers, he came through town, and Elliott sat and talked with Withers about his 1972 hit.

Sometimes performers went beyond the usual interview, really opening up a side of themselves typically guarded from the media. Bruce drew them out. One such interview was with Rawls. He told Elliott he was not having much success selling records in England and could not figure out why.

"The way the British announcers pronounced his name, it sounded like how they pronounce toilet paper. A *loo roll* won't help you sell records."

The ego that served Bruce well from his WYEN years of 1971 to 1973 didn't always sit comfortably with everyone. He remembers a Smithers observation after Elliott finished a show and thought he had done a nice piece of radio.

"Ray took me to his office, and I figured he'd agree with me."

Elliott remembers Ray saying, "Bruce, just so you know, although it's great that you did a really good show, nobody cares. Now I care, and you care, but the office people, they don't care."

Elliott says Ray meant to keep his ego in check and reinforce that it's a team effort and not just "all about me."

Elliott has only good thoughts of his early years in radio and loved working for Ray. He readily points to all the great opportunities and how much fun he had at WYEN. But the reality of radio often intervenes in some form, and for Bruce, his midday show was beating WMAQ, and so as he points out, "WMAQ did the only rational thing they could and hired me away." Elliott moved on to the Merchandise Mart. He later worked at WISN in Milwaukee, WBAL-AM in Baltimore for nineteen years, and finally his current position as program director at WILM-AM in Wilmington, Delaware, and sister station WDOV in Dover.

"Elliott in the Morning is smart and funny at the same time," he quipped.

For a boy with a soprano voice, radio may not have been his first thought. Bruce started his assent into entertainment through dinner theater singing "So Long Farewell" as one of the Von Trapps in *The Sound of Music*.

"I earned some money for dates, but when my voice changed, I could no longer play children in theater … so my solution was to go out to a radio station whose call letters I can no longer remember. Why not try for a radio job? I lied about my age, but I got the job and began to work to pay for dates. I don't think I thought this through because dates usually happened on Saturday evening, and guess what shift you have if you are the rookie coming in? That Saturday evening shift did away with my dating life for a couple of years."

Elliott got so busy he'd work at WEAW in Evanston while attending college. But for a time, Bruce was working on several stations at the same time, using different air names.

"I couldn't remember what air name at what station unless I carried a piece of paper around with me with the call letters and my air name. I was working at WGLD, WOPA, and WEAW, and migrated completely to WYEN, then to WMAQ."

Today, Elliott supervises a news department of three to four news people, hosts a news talk show, and works twelve to fourteen hours a day. Bruce works for Clear Channel, the largest broadcast group in the country. He anticipates the possibility of doing something interesting with them somewhere down the road if possible.

"I'll be here in Wilmington for whatever time they want me here. If Clear Channel wants me up the chain, I'll be delighted to do that too."

Mike Drake

Mike did not start out as the WYEN engineer. Walters and Smithers had their friend Ivan Bukovsky from WFMF join them as the chief engineer of WYEN. After Bukovsky's death within the first year, Mike stepped in, dropping his radio name, Drake, and using his birth name, Roche.

From the Walters Heinlein collection.

Most of what I can tell you about Mike has more to do with him as the station's chief engineer, a position he was in at the time I came to WYEN in 1976. I've met and worked with at least a dozen chief engineers at various radio stations. I've told them about problems with recording equipment, and later in my career, I'd let them know about problems with my computer programs. These chief engineers have been typically brilliant. This can be intimidating. I'd explain a problem the best I could, and I suspect they'd often keep their laughter in check. Of all the chief engineers I've known, I'd say Roche was one of the easiest to talk to about equipment problems. He also became a good friend. We talked sometimes for a half hour or longer in the evening on the days he'd work late. I can truthfully say it became fairly difficult at times to go back into the WYEN newsroom. Mike was just too interesting.

John Zur

"You hated him because he's a good-looking guy, and he sounds great." —Bill Jurek

"John with his cap and preference for a Corvette captured the image of a Seals and Croft album." —Paul Brian

"John was mysterious on-air with a deep, resonant voice." —Rob Reynolds

"A ladies' guy on-the-air, John was very smooth and slick." —Wayne Allen

"Stew, come and sit," John said almost every evening as my shift ended. John and I commiserated. Any problems I had, I'd tell John, and he'd lean back in his chair and give me a thoughtful answer. I could never tell on his face which he enjoyed more, helping me or entertaining his listeners. He'd built a good-sized audience of men interested in his knowledge of sports and women in love with his deep tone. John's voice filled all the notes in a symphony of sound. Women requested songs and I'm not sure what else, but he never got embarrassed, and he knew how to respond and keep them listening. He just handled himself professionally.

Back on-air, John sat slightly forward, towering above the WYEN console, talking to his audience, moving slowly through the language, mesmerizing those who'd come in contact with his message ... and then he'd turn the microphone off and sit back in his chair for a little while.

Listening to Zur's show was either like making rice or playing ball, depending if the listener was male or female. Zur was the heat source, a blue flame, touching the round bottom of a pan, cooking the rice bubbling in steaming water. The cover was clear, and he could see into the mix of steam and rice, turning up the heat ever so slightly until steam blew out, whistling from an opening at the cover. Zur just as easily talked sports and made you feel as though you waited in the batter's box or had taken a snap from center, or raced through a high-banked turn toward the checkered flag.

At a party held by my friend, Jeff James, I heard something that brought me back thirty-six years. In his extensive music library in his basement office, Jeff picked a cassette tape labeled WYEN. He knew I'd been working on *The WYEN Experience*. The date on the cassette case was April 1975. Jeff popped the tape into a cassette deck, hit play, and a voice powered through each of his speakers. The voice was Zur's. I again could see him sitting in his chair, cap on his head, talking about whatever was important ... perfection for his 6 p.m. to midnight shift.

THE WYEN EXPERIENCE

Bruce Davis

Not exactly *Good Morning Vietnam*, but darn close for Davis, a photojournalist and radio reporter for the Army's Nike—Hercules missile sites in the Chicago area. Bruce found time to do radio on weekends (Saturday and Sunday, noon to 6 p.m.) and evenings. When he finished his enlistment in 1973, Davis worked in the announce booth at Channel 11. He asked to tell his own story for *The WYEN Experience* as one of the original deejays on WYEN when the northwest suburban station went on-air in late 1971.

I was serving in the army stationed in Ft. Sheridan and had completed gigs at WLNR in Lansing and WEXI in Arlington Heights where I worked with Pat Cassidy. Ray hired me to do record shows and interview guests to fulfill our public affairs requirement.

For the next few years, I split my time between the Chicago and Milwaukee markets before going into the financial planning business from which I recently retired.

I worked not only with Smithers, but with Jurek, Zur, Bruce Buckley, and Reynolds. I remember that we had loyal listeners in the northwest suburbs and on the north side of Chicago, but there were problems with the signal penetrating Chicago proper and beyond ... emanating from the tower site in Arlington Heights.

We were housed in a fairly new office complex on Devon Avenue just a little south and west of the Touhy 294 Plaza. I remember that Ray and a few other staff members put together some very clever promotional spots for the WYEN bumper sticker. One in particular featured Ray as Richard Burton plying Elizabeth Taylor with the coveted bumper sticker.

Public service duties led to a steady stream of memorable characters. I was once approached by a young spokeswoman for the Des Plaines Antivivisection League. I had to check a dictionary to see what this was all about. She sent some promotional material, and we arranged for her to arrive at the station during my Saturday afternoon program.

In hindsight, I should have pretaped those segments, but everything was done live. She arrived at the station, seemingly prepared for a fifteen-minute interview. She sat down, clamped on a pair of headphones, and patiently waited for the record to end. I did a time and weather check and then introduced her.

Her face went from passive to contorted terror in a nanosecond. She looked at me, and if you listened carefully, you could hear her say, "I ... can't." Panic City! I stared at my notes, explaining to my WYEN listeners what her local chapter did ... and then I played a record. While the record spun on the turntable, she assured me she had composed herself and was ready to continue with the interview.

The record ended, I reintroduced her. Then I asked her about some upcoming seminar being held by her group, and you guessed it, she froze again, only this time, she opened her mouth, but nothing came out. No sounds! I hadn't let go of my notes and read what the local chapter was doing, gave out the phone number and address, and thanked my guest for coming.

Although the pay was fairly low—$2.50 to $3.00 an hour—these were great times. The army was paying me and providing room and board while I had a ball playing radio. I was a twenty-four-year-old "kid" with an opinion on everything and not hesitant to express them. Off-air, I criticized the all-request format because it sounded to me like we had about a dozen records in the music library. That didn't go over well with Mrs. Walters, and from what I understood, she directed Ray to fire me. He was pretty good about it. Smithers gave me some severance hours and wished me luck. A little of Ray's luck must have worked because I became a booth announcer at WTTW where the union scale was $6.00 an hour, pretty good for 1972–73. I stayed in a part-time, fill-in role until the early 1980s.

I ran into Ed seven years later at a Radio Shack store I was managing in Waukesha, Wisconsin. He had just bought the local AM/FM (where I had worked about six months earlier) and was buying some equipment for the station.

Bill Jurek

Howard Miller, Joel Sebastian, and Clark Weber were stops along the way on Jurek's route of Chicago radio stations on off-days from high school. Bill wasn't content listening to these popular announcers on his transistor radio. Every chance Jurek had, he'd go with his cousin and watch announcers in person. He would start out at WLS AM and see Weber on-air, and he'd go over to WIND AM and watch the number-one-rated Miller show.

"This was back in the 1960s, and the funny thing was we'd get off the elevator on the third floor of the Wrigley Building, and the receptionist would say, 'You guys again. It must be a day off from school. You know where to go. Just don't get into any trouble.'

"We'd stand and watch Mr. Miller on-the-air."

From watching Miller and the record turners in his studio, Bill and his cousin went over to WXFM on Michigan and Wacker Drive and talked to the announcer and listened to beautiful music for a while. Sebastian and WCFL were next on his *radio-around-town tour*.

"Then we'd go to the Carbon and Carbide Building where we'd see Walters, general manager at WFMF on North Michigan Avenue. You had to take two elevators to get upstairs and walk up another flight of stairs to get to where the station was, and I got to know Ed."

In my phone interview with Bill, he expressed that he has "radio in his blood." I wasn't surprised, not the way Jurek described what he'd do every time he was off from school. Bill can remember his interest in radio started

THE WYEN EXPERIENCE

at the age of eight, but then his parents wouldn't let him go downtown until he was high-school age.

At the age of nineteen, Jurek attended the University of Wisconsin and worked on a small radio station in Ft. Atkinson. During Thanksgiving break his freshman year, Bill came home with a tape of his radio show and had it critiqued by Bob Longbons of WDHF.

"It was one of those why-you-should-hire-me type of things. Longbons sent back the tape with a nice critique, and I wanted to thank him by taking him out to lunch. He said we'd go to lunch, but said, 'I'll tell you what, I've got this copier. Why don't you go into the studio over there and record it,' and I said, 'Okay fine.' I sat down and did four or five takes. I told him I was done, and he came in and he said, 'Let's listen to it.'"

Bill suggested he stand out in the hallway and let Longbons listen by himself.

"Bob motioned for me to come back in where he had listened to the tape on an old Ampex reel to reel."

Longbons instructed, "Well, you need a little work on it, but how would you like to work here?"

"Excuse me, are you talking to me?"

"Yeah, I don't have much, just evenings—Monday through Friday, 6 p.m. to midnight—but I'm looking for somebody, and if you are interested, you have the job."

"Yeah, I'll take it!" Jurek blurted out.

"You're going to start on Christmas Day."

"No problem, I'll be here."

With all the excitement of his first Chicago radio job, he momentarily forgot he had to finish his semester in college before he could start a regular shift. Longbons told him not to worry. He promised to cover the shift until Bill finished the semester.

"So I worked at WDHF through until 1972, and the irony of it was the same guy who hired me, Bob Longbons, fired me because of a competition type of situation. I got my baptism under fire that day and vowed to never work for just one station anymore. I wanted to work for all the stations on the dial."

Jurek walked across the street and started working for WKFM in Chicago and WLTD in Evanston and found out WYEN FM was signing on as the newest station. Bill had already known Walters, so he went over to the Des Plaines-based station and got hired. Instead of a position as a full-time announcer, he started working for Ed in sales and worked on-air on weekends 6 p.m. to midnight. Bill says WYEN was an easy sale for clients mainly because Ray created spots that were as good as any agency spot.

"Everyone got along at WYEN. This wasn't a backstabbing kind of place. WYEN was unique in that it was a radio station that let you develop a personality. If you stumbled and stubbed your toe, that was okay. The overall sound of the radio station made up for it. WYEN was almost an instant success from the moment it went on-the-air. Ed deserved that because of all the trouble he had getting the station approved by the Federal Communications Commission. The success he enjoyed was well deserved because it couldn't happen to a nicer guy."

Around the time Smithers left WYEN, Jurek left too.

"You could go back and visit WYEN. They'd welcome you with open arms. I sat and talked for an hour or two and then left."

Bill moved on to RKO General with their oldies format and got into the WFYR culture, which took up a lot of his time. His goal was to make it to a big company like NBC, which he did as a staff announcer.

"Bob Pittman approved me working on country music station WMAQ-AM 670 at the time he was program director, and he later founded MTV. I've worked with very creative people in my years on radio."

Jurek said he was not limiting himself in broadcasting.

"I wanted to be a broadcaster, and when I say that, I enjoy the engineering side of the business, the sales side, and the announcing side. Jim Hill of NBC told me I'm a unique guy in what I've done because I sound like a newsman doing news, an announcer doing a music show, a talk show host doing talk and community affairs, and can easily slide into sounding like a staff announcer when it's time to do the station breaks on television. That's everything I've wanted to do."

Bill feels the opportunities he had at the beginning of his career don't exist anymore.

"WYEN and WEXI and similar stations gave young people a chance to get hired and started in broadcasting without going out of town. You could hone your skills and become more professional and then move on to downtown Chicago. The business is not the same today. I will say with this reservation that if I had to go into it today, I probably would not because it's not as much fun as it was, and the opportunities that were there before are not there anymore. It's all regimented, it's all automated, you only plug lines in here and there ... and the creativity on the local level is gone. Everything has to be homogenized today, where years ago, anything goes."

Jurek has worked in radio and television on dozens of stations for well over forty years and currently manages CRIS Radio (Chicagoland Radio Information Service) at the Chicago Lighthouse. He's also serving a two-year term on the Video Programming and Emergency Access Advisory Committee of the Federal Communications Commission.

Chapter Six

HOME SWEET HOME

"God Bless America"
While the storm clouds gather far across the sea,
Let us swear allegiance to a land that's free,
Let us all be grateful for a land that's fair,
As we raise our voices in a solemn prayer.
—Irving Berlin

"Are you ready?"

"Yeah, go ahead," Sergeant Mike Burns said, catching his breath, "serve."

"All right, four serving two!" I shouted, stepping forward, driving my racquet into the ball with a sweeping underhand.

Skokie's best zipped to the back right sideline. Burns caught up to the ball with the full reach of a man taller than six feet. He returned the shot without sprawling out on the racquetball court. I smashed the ball low against two walls, forcing Mike's backhand shot. Then I hit the killer. He backpedaled into the wall, bounced off, and stuck out his racquet, desperately trying for contact. My killer pushed his racquet into his chest, but he returned it ... weakly. The ball barely bounced off the wall. I rushed forward and slammed the ball only a few inches off the ground. It streaked across the wood floor. The sergeant ran almost through me in the center, just reaching the first bounce off the wall. He hit it without thought.

"Ah, crap!" I muttered, with my hand over my mouth only seconds after my frustration registered with him.

"Was I in your way?" he asked. "We could do the serve again."

"No, I should have moved faster. Go ahead and serve."

I put the ball in his open hand. He knew I'd held back from getting him to do the serve over. However, I did not normally play a police sergeant of a suburban department that may eventually contend with a neo-Nazi march in Skokie.

Racquetball was supposed to reduce my stress, but the job and the neo-Nazi period added exponentially. I played a heck of a lot of racquetball in the 1970s with Brenner and even later after WYEN with Mark Krieschen. He eventually became general manager of WGN 720AM. Jeff and Mark were excellent players. Burns and I played at least once a week for several months at the Morton Grove Community Center. But after a few months, the whole neo-Nazi threatened march in Skokie began drawing tremendous amounts of attention worldwide. Burns became the main spokesman for law enforcement. Yet inside those four walls of the racquetball court, Mike and I didn't talk about the threatened march, and I certainly didn't bring it up. The world's troubles stayed outside. He appreciated the professionalism and treated me equal to all other media. But the whole march took a toll on our racquetball, and our playing time dwindled over the months the proposed march dragged on. Racquetball was certainly secondary to the troubles in Skokie, and I eventually put away my racquet and never played Burns again. Instead, I focused on the troubles ahead and used my hotline news phone.

"May I speak with Frank Collin?"

"Wait a second, I'll get him."

One of Collin's storm troopers put down the phone. I waited. I knew Collin couldn't pass on an interview. The leader of the Illinois Nazis of the National Socialist Party of America attracted media attention for weeks after dropping Chicago for a Skokie march. The media had been asking for Collin's specific plans in Skokie. I sensed from what I'd seen, he believed his words fell on a breathless audience. Collin had threatened to march in Skokie, citing First Amendment free speech rights, backed by the American Civil Liberties Union, ACLU, after the court granted an injunction against a march in 1977. He deftly handled reporters for radio, television, and newspapers, however, the public chastised the media for handing over a forum for his anti-Semitic speeches. Collin made forty pro-Nazi speeches in Chicago from 1971 to 1975. By 1977, the neo-Nazi leader was shut out of Chicago.

During the height of the threatened march in Skokie, I felt such an obligation to WYEN listeners, I couldn't move away from this story.

I was still waiting for Collin on the line. I sensed in my interview and others he'd done with the media that he believed his interviews had a direct pipeline to the suburbs and Chicago, however, as a rule, no interviews aired live. This allowed broadcasters editorial and audio control over any excesses in an interview. I prepared to do the same.

I had not actually met Collin in a face-to-face interview, so I couldn't picture what he looked like until I saw him on television. He did not look intimidating. However, the brief video of him didn't bring any depth to his character or personality. I could tell he maintained a controlled presence in a media barrage of questions. People were livid that Collins and the neo-Nazis could not be stopped from rolling out their brand of hatred before the very eyes of those who've endured great losses.

"Hello," Collin said.

I paused, not sure what I'd say next. His "hello" sounded not unlike the salutation of a neighbor walking by your house with a dog on a leash. I realized I'd just jump in with the radio station call letters and get to the heart of my question. I couldn't imagine talking about the weather or Chicago baseball.

"This is WYEN Radio. May I ask a few questions about your plans for a march and tape it?"

Who was I kidding? He did not need a prompt as in an opening question.

"Okay," Collin gave his approval for taping the phone conversation.

I heard him take a big breath. Before I could mute the phone and turn on the tape recorder …

"I can prove that Jews and blacks are inferior and they're all subhuman …"

"Wait a minute, please. I'm missing what you're saying … wait."

I hardly got the reel on record. This wasn't going well.

"Have you changed your mind about Skokie?"

I hoped to get the subject off what he'd been saying.

"We'll march in Skokie. The law's behind us."

"Aren't you concerned about inciting a violent response?"

I had my tough question out there now. I scratched out two other fairly tough questions. I had taken enough of his time.

Before he could answer, I hoped he'd take a moment and rethink what he was doing. Yes, it's true I was either practicing wishful thinking or plainly deluding myself. But what if he changed his mind? Could it be he'd had enough from the media questioning his every step?

I expected vigorous protests from people unwilling to step aside despite unfavorable rulings from the court. I didn't mention to Collin that the heavily Jewish population living in Skokie would not sit still for this. He had to have known the danger; he probably didn't need a reminder. He also was keenly aware the police in Skokie had to protect the neo-Nazis from anyone and everyone. I'm talking about protecting Collin's neo-Nazis from people likely lined five and six deep on Lincoln Avenue.

A march in Skokie had the prospect of a greater impact than in Marquette Park because of the heavy concentration of Jewish people, especially Holocaust survivors in Skokie. They had numbers forever tattooed on their arms, and they weren't about to let neo-Nazis bring back the *hell on earth* they suffered. The village of Skokie passed ordinances preventing the neo-Nazis from wearing their uniforms and swastikas in a march, or spouting any derogatory words, killing any real chance the Nazis had of marching there … so … Collin felt he had no option but to turn to the ACLU and Jewish attorney David Goldberger in what came to be known as the Skokie Right to March case.

"I'm ready for Skokie; we're ready to march."

Collin ignored my question asking him whether a violent response in Skokie worried him.

"I'll tell you, Jews and Blacks are both inferior for four reasons …"

Enough already, enough. I wanted to shout!

I held the mouthpiece almost in my mouth, ready to ask again. Then I realized the phone was still on mute.

"Okay, I've got enough on tape."

I guess that's what the ACLU meant by protecting the neo-Nazis right to free speech and the signs they planned to carry, *White Free Speech.* The decisions Collin made for the neo-Nazis were news, but the philosophy of hatred was fairly clear from the start and did not need another rehashing.

My stories on WYEN were news stories, not editorials or commentary. I arguably remained objective and reported without bias as best I could. People needed information. A good reporter, though, questions motivation, looking for what's underlying someone's actions. What was Collins's motivation? Was his motivation strictly aimed at showing his right to free speech was disregarded, so he'd gain publicity by announcing a march in Skokie? Did he truly not want to march in Skokie?

The US Justice Department stepped in, and the Chicago Park District relented, suddenly allowing for the march near the neo-Nazis Marquette Park headquarters and a secondary demonstration in Lincolnwood. This decision unhinged Collins's Skokie plans.

While Collin tested his free-speech rights with strong support from the American Civil Liberties Union, the Jewish Defense League (JDL) behind Rabbi Meir Kahane brought another angle to the neo-Nazis planned Skokie march. I knew very little of the JDL and Kahane before the planned march became such an international story. Because WYEN studios were so close to O'Hare Airport, interviewing the rabbi at the airport seemed fairly easy. Moments after the JDL landed and filed into a waiting room, I approached the rabbi's entourage with my recorder and very visible flagged microphone. Which one was the rabbi? I wondered. He and what I assumed were his bodyguards and JDL officers formed a semicircle.

They were man giants—not only tall, but wide. Had they worn football uniforms, they'd pass for middle linebackers. Each member of the JDL dressed completely in black. I turned, studying them, measuring each man's physical toughness. One of the JDL stepped behind me, and now I faced him. He didn't smile. He didn't say a word immediately, just stared at me. The fierceness in his eyes made me glad I wasn't his enemy. He was already psyching me out. I felt his strength and determination. He was unrelenting and unmovable. This was Rabbi Kahane.

"I'd like to ask you a few questions about the proposed march in Skokie."

After interviewing Collin, I had to know the resolve of the JDL and the Holocaust survivors. I hadn't expected Kahane's answers, because I did not know from where he came.

"Sure, go ahead."

"Rabbi, why are you here at O'Hare Airport?" I asked, staring straight ahead, keeping focus only on the rabbi.

"I'm speaking in Skokie on July Fourth for a rally against the march."

Time for the meat of the interview, the question with the longest response, I thought.

"What would you say to the neo-Nazis that hope to march in Skokie?" I held my microphone at his mouth, though he turned briefly to the others, acknowledging them.

"We, the JDL, would not say anything to them. We will kill them; then we will bury them."

The starkness of his reality nearly pushed me into flight. But I stayed wondering whether he could truly mean this?

"Can ... another way be found?" I offered a forced smile.

Without blinking, without moving his head or body, the rabbi stared at me.

With the absolute unchangeable law of the JDL, Kahane repeated, "We will kill them; then we will bury them."

Done! I could not ask anything more. I stumbled through my final question because his direct answer caught me off guard. Kahane had no intention of backing down or modifying his position. Looking at the faces of the JDL, I'm quite sure they'd follow through on every word Kahane spoke.

As he had promised, Kahane spoke in Skokie before a large, enthusiastic crowd.

"I give you my word. The obscenities that sit in the office in Chicago will no longer sit there. The Jew will never again allow a cancer to emerge and swallow us up."

To that, the crowd chanted, "Never again."

The Skokie Police Department and other neighboring police departments were not passive onlookers in this developing drama. Police officers knew their role, whether they recognized the irony or not. They'd have to literally protect the neo-Nazis from the masses.

Pretty much every week something new popped on television, radio, and newspaper reports from the neo-Nazi leader or Skokie's village president. The court also cranked out news. Seemed a hate group's First Amendment rights were preserved at the expense of a fair number of Skokie residents who themselves were targeted for extermination and then were threatened with the prospect of watching neo-Nazis march in uniform in front of them with full protection from the very police whose salaries were paid by the victims.

WYEN News stood in the middle of this whole issue. We were the suburban station, and this international story unfolded at our doorstep. We were given a great responsibility. I'd stay with this in every way I could. The First Amendment issues, the fear of Nazis, the Holocaust survivors, and the village of Skokie and their fight to prevent a march represented in its totality one of the most significant stories I knew I'd ever cover in my career. I truly believed people would have died that day in downtown Skokie, even with hundreds of police officers and Illinois National Guardsmen on duty and a perimeter of a thousand members of law enforcement. Losing loved ones in the Holocaust hardened the survivors into an unshakable vow. They weren't about to let hate reach into their homes again.

The world watched and prayed for a miracle resolution ... a resolution that arrived before any blood was shed. Word spread fast. No march in Skokie! It took a while for five thousand village residents of Skokie, each a Holocaust survivor, to recognize they were safe in their own community again. Forty-thousand Jews living in Skokie felt the incredible stress leave their bodies.

Because the court challenges dragged on, I ran through my mind what might happen in Skokie if the neo-Nazis marched. I had to prepare myself.

Thousands of people held hands and sang in one powerfully loud voice:

God bless America,

Land that I love.

Stand beside her, and guide her

Through the night with a light from above.

THE WYEN EXPERIENCE

I saw myself pulling my recorder out of my bag, recording the singing of "God Bless America" and ad-libbing my impressions of this historic moment.

"On this date, they won't be quiet. They'll sing of their great love of America, holding hands, symbolizing strength as a chain unbroken, and explaining with their voices as clearly as possible:

God bless America, My home sweet home
God bless America, My home sweet home.

Just as fast as the last notes of "God Bless America" faded in a melody of peaceful resistance, police escorted a small group of neo-Nazis, their marching boots and uniforms instantly visible. The sound level rose and transformed into a cacophony of screams and cries, ratcheting louder by each step of the storm troopers. The sweet music of "God Bless America" had held promise ... but the new song of suffering, anguish, and disgust held only the promise of physical retribution. Protected from physical assault from bullets, rocks, and bottles, a thousand police officers were not prepared for the sound. Unlike Homer's writing of how Ulysses protected his men from the siren's song in the *Illiad*, none of the officers in riot gear had trained for the unleashing of an emotional tumult. Yet they endured impending disaster, and fortunately for them, the neo-Nazis did not march far, ducking behind a building for a quick escort out of town.

The reality is Skokie streets never accommodated the march of hatred—though the town would have, but not by choice. Skokie streets never accommodated the Holocaust survivors—though the town would have tearfully. Skokie streets never accommodated reporters—though we would have captured the moment for the world. Skokie streets never accommodated the police—though the town knew how absolutely necessary they were even as they protected brown-shirt marchers. Instead, Skokie streets accommodated what Skokie streets should accommodate then and now and always. Peace.

Today, schools teach the Holocaust. Teachers read books on the Holocaust, and they talk about intolerance and how you must fight hatred and indifference. Field trips to the Illinois Holocaust Museum and Education Center in Skokie have given thousands upon thousands of students a direct account of the lives of people that endured hell in the death camps and many others that succumbed to a world without humanity.

For some, this was a test of whether to set limits on free-speech rights.

For the Holocaust survivors and the memory of their loved ones long since killed in the death camps, this chapter, Home Sweet Home, reminds us never to accept a climate that opens the door to hatred.

Chapter Seven
MUSIC IS MY FRIEND

Make the most of music magic,
Hear it clear across the lake.
We've got musical things to say.
O'Hare Airport, the CTA.
From Milwaukee to Gary,
Listen to Request Radio,
WYEN
WYEN ... WYEN
The bright exciting sound of today!

Reynolds's Rap

High Rolling Rob Reynolds knew this WYEN jingle by heart.

"I know every one of them," Rob emphasized. "I played them on-air often."

The jingle package was originally for KBIG from the 1960s. In my interview at Rob's home, he told me Smithers adapted them for WYEN.

Reynolds, Zur, Allen, Stouffer, and other WYEN announcers enjoyed playing the jingles as much as listeners enjoyed hearing them on-air. I told Rob I thought the jingles really drove the character of the station with its very catchy sound, but I wanted to hear what Rob had to say for his six years at WYEN.

"I have many memories of Walters and the entire crew ... most of them very good. Those that weren't, I have learned, ultimately shaped me into a better businessman and person."

Radio sparked his interest as an eight-year-old. Rob called himself a radio groupie. His interest never wavered, just grew stronger, and by the end of his junior year at New Trier East High School, Rob counted on a promotion at the school station. He believed he'd be named station manager of the 33-watt WNTH. However, the high school station's management denied him a promotion.

Rob in turn justified in his mind, "I will show them!" Rob applied for a weekend shift at WYEN.

"I walked into the station (no one ever asked how old I was, I could have passed for my early twenties, I guess), and Smithers, the program director, took me into the production studio and started rolling a reel of tape."

Ray cued a record on the turntable.

"He asked me to play disc jockey, and that's what I did."

Rob's confidence showed in his performance but stopped short of figuring what Smithers felt ... but Ray didn't take too long.

"During his newsbreak, Ray offered me a job on the spot that Friday afternoon and told me I'd start the next morning at six o'clock."

Just a teenager, Rob passed the one hurdle, landing a job on a professional radio station. *Now for the second hurdle*, he thought.

"Smithers insisted I'd need a radio name. Brumbaugh just didn't work. I looked around the room and saw tin foil on the level-devil or whatever they called it, and I mumbled something about Reynolds wrap."

Momentarily undecided, Smithers switched on the microphone and blasted, "Tomorrow morning at six, it's the all-new Rob Reynolds show right here on Request Radio. And that was that!" exclaimed Rob. "At sixteen, I started on-the-air at WYEN. It wasn't five months later when Smithers called asking me to fill in for Bruce Elliott, the midday guy."

Rob knew he couldn't fill in, but how does he tell the program director?

"I told Ray I couldn't miss my classes, and he said, 'Oh, college students miss class all the time.'"

A moment's pause, Rob looked directly at Ray and corrected him, "High school students don't miss class."

Ray was dumbfounded, remembers Rob, but that became a defining part of his career.

"Eighteen months later, while a sophomore at Northwestern, I was promoted to the afternoon drive gig, hence *rolling you home*, and became the music director after Ray left."

Despite what Rob considered Walters's frugal nature, he said with total confidence he had the best college job of anyone he knew.

REYNOLDS IS READY TO STEP OUT OF THE STUDIO FOR FIVE MINUTES OF NEWS.

Rob emphasized the years he worked at WYEN were the golden years of Request Radio and stressed the talent pool was incredible. I asked Rob about some of his coworkers.

"Smithers was an absolute genius in programming and music selection."

Rob remembers Stouffer as a great announcer, and he said he attended her wedding. Jayne Netches, another friend, had a late-night shift and later became big at A&M Records. They'd go out for breakfast after her shift. He remembers Dan Diamond as a slick announcer and that he always knew what he was doing.

Meier was then and still is one of Rob's best friends.

"As a straight music deejay, however, we all voted him most likely to return to pharmacy school."

Rob could tease, but then so could Garry. Rob recalled that meaningless banter between songs simply wasn't Garry's thing.

"There is a certain irony that I was offered a weekend job at the then brand-new WLUP though Devine was working weekends there. I remember asking what it paid. I kid you not, the program director told me five dollars an hour. I remember thinking I can't park my car at the Hancock Center for what I'd make on a four-hour shift."

Rob stood his ground despite the program director challenging him, "Do you want to stay in radio or not?"

Rob turned down the job, but he continued, "Meier wound up working there all night before the brand-new Steve Dahl Morning Show. That's how Garry's rise to fame began. I am guessing he was making five bucks an hour too."

Rob was all business at WYEN, and he looked to me like one of the original multitaskers. He would play music, talk on-air, take request calls, talk to record promoters, and write commercial copy, all during his shift. He had some pressure to cut back on the multitasking, but Rob could not.

"I was music director, and I was going to maximize what I'd get out of WYEN. We were a Chicago reporting station for *Radio and Records Magazine* and the *Gavin Report*. As a result, since there were no adult contemporary stations in Chicago, people looked to what WYEN was adding to the playlist. I went to dinner with record promoters where a bottle of wine cost more than I made in a week."

Rob was doing commercial work on the side.

His snappy read, "Evening Tides Waterbeds, putting more of Chicagoland to sleep every night," emphasized he had a way of internalizing his clients' messages that he'd never forget, nor would listeners.

When Rob was let go by Walters, he went to the advertisers, saying he'd like to keep working with them as their voice. They wanted more. This was the moment in Rob's career that would determine the direction for the rest of his life.

Rob responded to a question posed to him by an advertiser, "Could you do newspaper ads?"

"I said yes, though I had no idea, but a friend of mine was a graphic designer and would help do the ads, so instead of producing local radio commercials, I became an advertising agency. As I would tell Ed many years later, with all sincerity, he did me a huge favor, as I would never have started my ad agency had he not pulled the trigger."

Reynolds owns Omnibus Advertising and manages a staff of twenty-five employees. He's also been the main voice of Speedway.

Musical Homage

Not just as the author, but as a WYEN employee and as a listener, I can't separate 1970s music on WYEN from Reynolds's work as music director, nor can I separate seventies music from Wayne Allen, the program director of WYEN during the disco craze. *The WYEN Experience* pays homage to the seventies music; Rob placed the music carefully on Request Radio's turntables for thousands of listeners, including me.

I've asked several former WYEN announcers to help me pay homage to the seventies decade since the music fueled Request Radio. I'll start with the homage since the 1970s was the key decade in my life, taking me from high school, through college, and into my first and second professional jobs at radio stations in the Chicago area.

The 1970s music and milestones in my life worked together to help me remember my first job as a busboy in 1970, graduation from high school in 1972, graduation from college in 1976, first full-time professional radio job at WYEN in 1976, and professional association with legendary WBBM radio broadcaster Mal Bellairs, owner of Lake Valley Broadcasting, who brought me to beautiful Crystal Lake, Illinois, in 1979.

Freda Payne's "Band of Gold" brings back wonderful memories from 1971 of flying on an airplane for the first time and visiting California for the first time. I'm sitting in a little outdoor café in Disneyland with my parents, and suddenly a small square of the floor opens, and Payne rises on a platform through the open floor and begins singing her hit. What an unexpected pleasure. That same summer, I started my first job as a busboy at Oscars Fine Food in Morton Grove. I was seventeen and in love with Colleen, a twenty-two-year-old waitress. The Partridge Family's "I Think I Love You" grabbed my attention. In 1972, I graduated from Maine East. Don McLean's "American Pie," out in 1971, was a big hit on the charts and was one of those songs that kids sang to themselves, at least parts of the song they could remember. I only remembered the refrain and sang about "them good ol' boys were drinking whiskey and rye" a lot. In 1976, I graduated from Southern Illinois University with a degree in Communication and had my first serious relationship breakup—the same year for "Don't Go Breaking My Heart" by Elton John and Kiki Dee, "50 Ways to Leave Your Lover" by Paul Simon, and "Silly Love Songs" by Paul McCartney and Wings. Also in 1976, August 16 (one year before Elvis died), I started working at my first full-time radio job at WYEN. A big Four Seasons fan, I remember late in 1975 through 1976 singing over and over "December, 1963 (Oh, What a Night)." In 1979, I moved north into McHenry County, just south of the Wisconsin border, where Chester Gould, the creator of Dick Tracy, lived. On the radio, Blondie performed "Heart of Glass." I started working on WIVS AM, the station purchased by the Bellairs family.

Read the comments on YouTube on songs and artists from the 1970s and you'll most likely read "this *was* music." I asked former WYEN announcers to think back to their years on Request Radio and tell me their memories of the music.

THE WYEN EXPERIENCE

WYEN Program Director Ray Smithers: Although recorded in 1968, "Is That All There Is" by Dan Daniels, Peggy Lee made the song a hit in 1969, and the song played throughout the seventies and was one of the most requested songs on WYEN. My dear friend and coworker, WYEN Chief Engineer Ivan Bukofsky, worked with Ed and me at WFMF. Ivan would always come into the control room when I played the song. He'd become a bit depressed lamenting the truth about the song's lyrics.

One afternoon, Ivan seemed very depressed and quiet. He came into the control room during my shift and asked if I would play "Is That All There Is." I did. That same evening, just after midnight, I received a phone call from Carol that Ivan had passed away from a heart attack. The entire staff came to his wake. I felt numb. The next day on my drive shift, I cued the song "Is That All There Is," opened the microphone, and said, "This is for my great friend and our Chief Engineer Ivan Bukofsky, who we lost last Friday and ..." I couldn't get anything else out. I had to play records for ten minutes straight without introducing the music because I just could not talk. "Is That All There Is" will always be Ivan's song, and the lyrics will always be true.

WYEN announcer Mike Tanner: I think the Rhino Record Company had it right when they released a multidisc collection of favorite music from the 1970s and titled it *Super Hits of the 70s, Have a Nice Day*. That decade introduced us to a wide genre of artists from Elton John and his debut hit "Your Song" in 1971 to Donna Summer, spawning the disco era. If we weren't introduced to John, would we have ever known or cared that today's hit singer Lady Gaga is the godmother of his child? Who can understand the lyrics in the hits that are on the air today? The lyrics were so distinct and memorable back in the seventies. One of our frequent requests was for "Operator" by Jim Croce, and those lyrics had me singing along in the studio. Who would have thought that while I was playing "Dancing Queen" and "The Name of the Game" by ABBA on my afternoon show, thirty years later that music would be turned into a Broadway musical. While I finally had the opportunity to program a station in 1984, the station manager referred to seventies music as a dud and would prefer that it not be played on the station. Maybe he just didn't like the Bay City Rollers or Andy Kim. (Actually, my least favorite song from that era was Debby Boone's "You Light Up My Life.") Sure there was a lot of "bubble gum" music during that decade, but how great it was to be able to play hits from the local groups like Chicago and REO Speedwagon. I ultimately got the station manager to realize the music of the 1970s was instrumental for creating new music radio stations like adult contemporary and dance.

At Request Radio, the variety of music that we played was as fun and memorable as the decade itself, including the Carpenters, Bread, and The Bee Gees.

WYEN announcer Garry Meier: You have to keep this in the era it's in. I met Burt Bacharach, Melissa Manchester, Herb Alpert, and Bobby Darin. They came into the radio station. They were the people well known to that era. We had a good parade of celebrities.

WYEN Program Director Wayne Allen: The music of the 1970s was true music ... enjoyable to listen to. Rock and roll was the real deal, unlike today's hip-hop and rap. You could truly enjoy melodic tones and not a lot of yelling. Call me crazy and call me getting old, but I loved the songs of the 1970s; Eagles with "Take It Easy," Richard Harris with "McArthur Park," and Aliotta Haynes and Jeremia with "Lake Shore Drive," and on and on. There was substance to the music of the 1970s. Country music was and still is storytelling. Rock had its beat, and the love songs were always the most requested. I compared WYEN to WLIT as formats were quite similar, yet at FM 107, we were more personable and had fun. Looking back to those days, I truly miss what it was like being on-the-air, and I miss my listeners. They'd call in their requests at 591–1166, a number that will be lodged in my brain till death do us part. The famous request line with the same listeners calling every day asking for us to play a song that they could feel close to ... a song that had the true meaning of music, not words.

WYEN announcer Roger Leyden: I was truly moved by John Lennon's "Imagine."

As the 1960s ended, with good and bad, it left a divided country, the generation gap, the Vietnam War, and don't trust anyone over thirty. The 1960s turned to the good with the landing on the moon and saw the early 1970s trying to take over on the good side of humanity and life.

In the fall of 1971, we hailed an anthem for all of humanity—"Imagine" by John Lennon. The soft, haunting piano intro invited you to step into the song and feel Lennon's feelings and thoughts. *Imagine all the people living life in peace.*

John Lennon never tried to force people to change, and I truly felt he believed it was our choice to seek out the good in the world. He was the genius pied piper showing us the right road to take, if that's what we wanted. "Imagine" was the starting point on the road to peace in the early 1970s.

Lennon spoke and sang his mind, never worrying how hard the critics came down on him. They did come down hard. For that alone, I hope you were on that journey of peace. I was, and I've never regretted one moment. Thank you, John Lennon, for making us think and for giving us a moment of pause. It was my pleasure to play "Imagine" on WYEN.

WYEN announcer and Program Director Kenn Harris: Long before I began working at WYEN, I was a listener. What I liked in particular about the radio station was the interesting selection of music. Mixed in with current hits and the familiar oldies were songs that had no chart history. Sometimes they were album cuts from artists that fit the format and already had a single on the charts, and sometimes they were just album cuts from albums that Rob just thought would sound good on-the-air. When I began working at WYEN in early 1978, I had an endless list of music questions for Rob. I was anxious to play those songs that I'd been listening to for several years. When I took over as music director, I tried to follow Reynolds's philosophy. He felt it wasn't enough to just play the single off a charted album. It meant listening to every cut off the album to find a song at least as good as the one released as a single. This meant listening to albums that had no chart success and were for the most part totally unknown. But in so many cases, there were some hidden gems on those records, and they played a big part in the unique sound of WYEN. Thirty-three years later, many of those songs are still my favorites and have made the transition from the open reel music tapes I made there, to cassettes and now to CD and iPod. Evening jock, Mountain Man Bob Walker, as we referred to him, called those unknown songs *Chinese aviators*. I once asked him how he came up with that term, so he asked me to name a Chinese aviator. I thought about it for a minute. Famous English, German, American, and French aviators' names came to mind, but I could not name a single Chinese aviator. He just nodded his head and said, "There you go." Whatever you want to call them, great artists like the Faragher Brothers, Gene Cotton, James Lee Stanley, Cecelio & Kapono, Megan McDonough, and many others found an audience on WYEN.

Serious fan and announcer Jeff James: I not only listened to Request Radio as a child, but I taped the announcers, the music, and the jingles. Ever since my parents bought me a record player when I was in kindergarten, music has been my favorite pastime. I grew up in Schiller Park, Illinois, and could hear WYEN clearly and other stations that drew my attention—Fire Radio WFYR Chicago, WDHF, WEXI, and WLOO. That's quite a spectrum of music. Listening to these stations and more on my transistor radio became a full-time hobby. But my best memory was when I called into WYEN for requests and heard my song on Stereo 107. One such day in the fall of 1972, I called during midday and got my request played, "Stir It Up" by Johnny Nash, and the jock mentioned my name. From that point, I pursued the dream job of working in radio.

I've worked as an announcer on WZSR-FM and switched to its sister station WWYW-FM, Y 103.9, playing music from the seventies and eighties. My special show was Saturday Night Live at the seventies on Y 103.9FM. I've enjoyed keeping the memories alive by not only playing the great forty-fives of the 1970s, but occasionally playing the classic Chicagoland radio jingles, including those classic big-brass sounds from WYEN.

CHAPTER EIGHT

TECH NO SAVVY SEVENTIES

I have in the WZSR-FM, WWYW-FM newsroom today amazing digital technology. One computer called NexGen can store an endless supply of music, spots, jingles, and any other sounds needed for a radio station. In the analogue world of the 1970s, we'd need several WYEN studios filled with carts, cassettes, reels, and records to hold maybe half the amount of audio that one computer can hold in the digital world. Each studio at STAR 105.5FM and Y 103.9FM has a NexGen network all linked together, so when the production studio produces a spot, or the news department produces a breaking story, public service announcement, or pretaped public affairs show, I can hear the piece in the newsroom, two on-air studios, production studio, and voice-tracking studio through NexGen. I can see the computer log of everything that'll play on a given day, whether it's today, tomorrow, or the end of the week. I can produce a public affairs show on a Monday and have it play at a selected time on Sunday morning while I'm home eating breakfast. The production director may ask me to produce a thirty-second spot for an affiliated radio station one hundred miles or one thousand miles away; distance doesn't matter. I can produce the spot in Adobe Edit on my computer and MP3 the spot to the production director. He can then mix the vocal with music and send the finished spot to the station one hundred miles away. A few minutes later, an e-mail from the station's production director says the spot is good and we'll see an e-mail attachment with the spot narrated and mixed with music. I'll burn a copy on CD or keep the audio in a file, and I'll have the copy for my records. We can accomplish all this in three or four minutes if we work fast. For speed and quality, no technology so far beats digital. In some ways, I miss the hands-on grease pencil, razor-blade kind of day, cueing records, rewinding reels of tape and then threading them, or popping in cassette tapes or searching for the white out for the typewriter. You can truly say you worked a full day at WYEN and other 1970s stations if you had white out on your hands, carbon paper on your fingers and shirt, and a few cuts from the razor blade you used for cutting and splicing tape.

I can't imagine how anyone in a 1970s analogue newsroom such as I described with reels of tape and cart machines could foresee what radio newsrooms might look like today. The changes in the 1970s seemed to me to move faster than the 1960s broadcast changes, but with every successive decade, improvements in broadcast technology seemed to rush ahead until now where most of the changes involve the social networks and website development.

I admit I had trouble keeping pace with technological advances. My sons, Brenden and Brant, thought my knowledge and use of computers was quite primitive, but also I was a dinosaur with iPods and iPads and the newest cell phones and Twitter. The boys referred to me as "Dad Before Time" or "Dad in the Dark Ages." This is just my opinion, but I believe young people think life as we know it began with someone carrying a Blackberry for the first time. Should they run across someone happily describing life without cell phones and computers, that person might as well be a Neanderthal.

I can e-mail, send messages on Facebook and LinkedIn, send stuff MP3, edit on Adobe Edit, and retrieve messages and texts on my cell phone. Isn't this good enough? Why would I need to use an iPod, iPad, or Twitter or some other new technology they seem to like? Maybe I'm hesitant because I'm still being confronted by new technology that I'd have to learn with all the other things I've already had to learn to keep up. I believe it's tougher for someone not born in the Computer Age to stay afloat with this rapidly advancing technology. I grudgingly admit I recognize I'm running backward in this race with technology.

"Dad, I can teach you how to use these things," my sons would say of an iPod or Facebook, or something more advanced on my cell phone. Sure, they know I can't even send pictures over the Internet without massive amounts of help, and I can barely troubleshoot my television cable, but then who really has an understanding of their television cable or dish system?

Someday technology will advance so rapidly that Brenden and Brant won't understand how to use it without asking their children. I'll have the last laugh then. Still, the incongruity in my life screams because I use digital technology, yet I'm not interested in iPod and iPad technology or apps.

Well, I'm not yet interested. I take longer to recognize the importance of new technology because I'm simply a "technical Neanderthal." I miss how technology wasn't so suffocating in the 1970s compared to today, where on many occasions, I've seen everyone in an office sitting with their cell phone or other technical equipment and fully immersed in texting, talking on the cell, playing games, or checking apps.

Secretly, I miss those WYEN days in the 1970s of tape splicing on an editing block where I had marked with a grease pencil and put the two pieces of tape together with editing tape, and then I ran the edit through the play head on a reel to reel. This must be the way someone feels driving a Corvette with a stick shift or way earlier, cranking the starter on a Model T Ford. This is driving! I really was building a story by splicing, editing, and taping.

Technology has given us iPods, CDs, and MP3 players, but also long-playing records on turntables. Yes the vinyl is technology from days gone by, but playing music on records was the way radio stations operated during the WYEN years, long before and some years after. The announcers cued a record before one of the WYEN news announcers sat at the console and prepared for a newscast. At the end of the news, we'd read weather and return to the announcer by saying a standard phrase and then start the turntable as we said our name and station ... and the motor would bring the song to the right speed and play the music.

I asked **Garry Meier** about the equipment in the WYEN on-air studio.

Walters didn't own the most professional broadcast equipment. We were running tapes from a cassette player while most stations then had standard four-track cartridge tape, specifically for broadcasting. It was the four-track version of eight track. They had different lengths for commercials and songs. This was the easiest way to do things in 1970s broadcast technology. Stations had hundreds if not thousands of these cartridges, many for "carting" songs (putting songs on the cart). You had these cart machines firing one cart after another. Walters finally broke down and bought one cartridge machine, and we thought we won the lottery because we had a cart machine.

To show the evolution of this business technologically, when I was at WCKG a few years ago, the only four-track cart machine I saw was holding the door so you could go to the bathroom and the door wouldn't lock behind. It was a door stopper, and here's an item that was as golden to radio at one time as oxygen is to us today. You think something will be golden forever, like the electric typewriter, and then it becomes extinct.

Bruce Elliott, an original announcer on WYEN, told me his experience with technology over the years.

Sitting at my desk, I can open an edit session on the computer and edit something I recorded yesterday. It's all internal to the computer. This is so much different from sitting with tiny little pieces of tape and trying to make sense of them. The editing process is so much easier with people working in music radio. They frequently voice track what they do, and that's a very different skill and would have been unheard of at the time of Clint Eastwood's 1971 movie, Play Misty for Me, *where he recorded his radio show on huge reels of tape. Today, you'd put down huge amounts of music on a hard drive and let it go at that.*

Dave Dybas, a WYEN engineer on contract from the mid-1980s, told me how technology was still bound to the mechanical world in the seventies and eighties.

Tape machines had turning motors and moving parts that required regular attention. These mechanical parts required attention and thus a full-time engineer to service them. Remember cleaning the pinch roller? Today's equipment has very few, if any, mechanical parts. The electronics are all contained in a few chips that are so small that they're nearly impossible to unsolder or replace. The playback gear is basically throw-away gear. Should the equipment break, you throw it away and get a new one. And, of course, much of the gear has been replaced by computers. Manufacturers have maintained a constant pace, creating equipment that requires little attention. This has led in part to the demise of full-time engineers.

Dybas is currently owner and operator of Sparks Broadcast Service and has served as chief engineer for Salem Communications-Chicago.

Meier, Elliott, and Dybas certainly know the differences in broadcast equipment over the years, having lived and worked through the last four decades of technological changes. This trip of technical advancement has been nothing but amazing in broadcasting. I've witnessed so much of this change that the story I'm telling you now seems so foreign in today's broadcast world.

The chemical plant in Skokie stood as a giant in the downtown area of the village. Searle's main plant had caught fire. Spectacular fires like this were not common in the suburbs, making this a huge story for Chicago TV news stations and radio stations. I already knew Wilkinson would call me, so I got ready, gathered my equipment, and rushed out to Skokie after he called. As I mentioned earlier, in those days, reporters did not have cell phones. I wanted to use the pay phone in front of the shops near Searle, but a reporter from WIND AM was hogging the phone. I listened to his report, and he was smooth and descriptive and had this really deep, distinct voice. He just wouldn't get off the phone, and I thought maybe I could go into one of the shops, but I just didn't have the guts to step out of my shell and see if a shop owner would let me borrow their business phone. So by the time I finally got into the pay phone booth, an hour after arriving, Wilkinson was pretty upset with me. He said they had to use information from others on the fire and explosion despite having a reporter—namely me—at the site.

Bill succeeded in making me upset with myself for failing to do what we're supposed to do in stories like this. I vowed to do a better job of covering on-the-scene reporting.

The best radio reporter in my experience in the Chicago area is Bob Roberts of WBBM 780. This is my opinion based on having worked with him and having listened to his reports for several decades. I don't know whether Bob had the same kinds of problems I faced in his early career, but I can truthfully say he is a broadcaster's professional ideal. You are about to read Bob's story about his early career at WYEN and beyond.

Chapter Nine

BEST RADIO REPORTER

I'm Bob Roberts.

    I glanced around the room. I was nervous, and with good reason. The clock on the wall of the nearly pristine, virtually new studios of Metromedia owned WMET on Chicago's Magnificent Mile read 10:30. Chicago Tribune and Sun-Times next to the phone bore the date April 4, 1978. Less than an hour before, Program Director Gary Price ushered me into his office and sat me down. I wanted to tell him about a blood drive story we'd done exclusively for a young listener and how the donations had exceeded what she needed. Instead, Gary told me, "This is your last day at WMET."

    No explanations were given, but in the past thirteen months, Price's ideas on what constituted a good newscast had changed by the day, if not the hour. What worked one day wasn't good the next. Dave Alpert had been shown the door in February. This time there was no buffer, no way to talk it through and no way back. I was done, and Randy Thomas henceforth would be the news voice of WMET.

    Days earlier, I had tacked up the framed certificate from the Radio-Television News Directors Association proclaiming me a full member. That was going home. But what did I do next?

    I was a month shy of my twenty-sixth birthday, I was out of a job, and I was scared of the unknown. A sick, grinding feeling emanated from my stomach. A jumble of thoughts raced through my mind. Did I screw up, or was I just the arbitrary victim of a program director who couldn't make up his mind? Was my work any good? Would anyone want to hire me? How would I pay the bills at the end of the month? Little made sense.

After staring at the phone for what seemed like an eternity, I placed one call and received another, which together would help determine my next three decades as a journalist. The call I placed was to WYEN News Director Stew Cohen. Stew expected me to cover a couple of stories downtown for him that morning. WMET and WYEN regularly shared stories and tape, a legacy of Dave's tenure at both stations. Randy wouldn't be gathering tape, that day or at any time during his brief stint with WMET.

Stew seemed as surprised as I was. Working for Price was never easy, but neither of us thought that I'd be shown the door so quickly after my promotion to news director. Stew mentioned to others at WYEN what was up as we kept talking. A few moments later, he indicated that Mr. Walters wanted to take the phone.

"Hello, Bob," he said.

I didn't know Walters and had never spoken with him. Nor had I expected to speak with him that day.

"I'd like to discuss with you the possibility of working for us," he said.

My ears perked up.

"We can't give you much," he said. "I can offer one hundred dollars a week, but you would be anchoring news during the morning show."

I detested working in the morning, but as a matter of survival, I wasn't going to say no. The Associated Press Radio Network and the Illinois News Network both counted on me for Chicago material, which was a good supplement to my pay, but impossible to do without direction; working for WYEN would give me access to the City News Bureau daybook, the Holy Grail for Chicago assignment editors. I had put together that daybook five nights a week less than two years previously, while CNB's overnight editor.

Until then, I'd managed to bypass suburban radio. I had gone from Bloomington, Indiana, and college radio at Indiana University, to CNB, and from there directly to WMET, working both as a WMET deejay and City News reporter simultaneously for nearly five months. Maybe I'd reached the number-two market too quickly. I certainly didn't know the diplomacy required of a manager. WYEN was a good place to land. I could begin packing my stuff into a box, saying my good-byes at WMET, and loading the box into the green 1975 Buick LeSabre convertible that I'd bought the day before with three thousand dollars cash. While doing that, the WMET newsroom phone rang. It was Rich Reiman, the news director of the two NBC-owned radio stations in Chicago, WMAQ and WKQX. I'd gotten to know Rich just before he was promoted, chatting at length with him while covering stories.

Bad news travels fast.

THE WYEN EXPERIENCE

"I heard you've gotten the axe," Reiman said. "Welcome to the club. Everyone in radio gets fired." It was hardly reassuring. I'd applied to WMAQ when WKQX had been an affiliate of NBC's short-lived News and Information Service under the call letters WNIS, but had messed up an interview for a job that had been mine to lose. I'd told Rich how bad I felt about that and hoped someday I'd get another chance. That day had come, but I was at the end of the bench.

"I don't have a lot of work. In fact, right now, you would work one day every two weeks, Saturday mornings, on 'KQX, because the regular guy (Steve Tom, a former reporter/anchor for WMET predecessor WDHF) wants the day off," he said.

Wonderful! It all fit into my schedule, and the WKQX pay easily eclipsed what I earned for the week at WYEN. I told him yes.

It quickly became apparent that I had a dilemma, though. Would WMAQ be happy I was working during the week in Des Plaines, at a station that most of the market could hear? For that matter, would the Associated Press, AP, be happy that I was working for WYEN, which was a United Press International, UPI Audio affiliate?

I told Stew my problem and said I had an idea. Why not use the air name I used on the radio at IU—Frank Rhodes?

Frank Rhodes it was, although it took WYEN morning man Wayne Allen a while to adjust. He'd go, "Hey, Bob," while promoting me during morning drive, and I'd have to shoot back at him, "Who's Bob?" or maybe, "The janitor left."

The WYEN newsroom was not what I'd grown accustomed to using at WMET. Phone interviews were recorded onto a consumer-grade reel-to-reel machine. The push-to-talk feature on the phone was difficult to use. Recording a report meant having to go to the production room, and I used the main studio microphone for my newscasts. But WYEN had the same teletypes as the big stations, and I wasn't shy about telling sources I had moved to the suburbs. WYEN was my home base, sometimes for many hours each day after 9:05 a.m., when my official workday ended, as I cranked out stories for AP Radio, INN, and occasionally UPI Audio, also using "Frank Rhodes" if I filed for UPI. It could get sticky if I filed something for NBC while not working for WMAQ, but I sometimes did so.

I liked working out of WYEN because it gave me a way to package my material to be shipped to the networks. Illinois News Network, INN, sometimes asked me to do its entire afternoon feed, including a pretaped newscast. That required the production studio.

I slipped into a grueling but workable regimen. I'd arrive about 4:45 a.m. to rip wire copy, check the newspapers, begin making calls, and prepare the morning newscasts. The first newscasts normally aired at 6 a.m., although if there was major breaking news and I was ready, I could go on the air at five, and sometimes did. For a part of 1978, the overnight slot was brokered to legendary Chicago rhythm and blues disk jockey Purvis Spann. When I did a 5 a.m. newscast, I was mindful that I had a somewhat different audience than Request Radio did during the day. I didn't get to know Purvis well but found him a fascinating individual—promoter, storyteller, and entertainer; no mere deejay.

Although I had been news director at WMET, that was never my job at WYEN, and I did not want it. That was Stew's job, and I was glad he was doing it. Stew and I worked well together, and Walters must have sensed that. Although he was never known for spending money, he hired two more people—Beth Kaye to do primarily rush-hour traffic, and Wally Gullick, who did news and sports. Beth would go on to work for WKQX's Murphy in the Morning show before becoming an educator, and Wally would become the great Walter Payton's publicist. But we were all twenty-something then, trying to get or keep a foothold in radio. The station promoted its news department, and it just spurred us on some more. WYEN Metro News was running on eight cylinders, and we were having fun.

My hours expanded as 1978 progressed at WKQX and WMAQ, and sometimes I'd have to take a morning off from 'YEN to do NBC fill-in, but Beth and Wally were quick learners. Everyone pulled their weight in the WYEN newsroom during the summer and fall of 1978. I'd joke with Beth, and Wally and I sometimes would spend several hours between newscasts gabbing about the Chicago sports scene, still managing to find time to gather plenty of local news. Everyone critiqued everyone else's work, and we all came away better for it.

If the story had regional or national interest, I'd be on it to try to make network money. We didn't forget where we were, though. While we battled the downtown stations, we were in Des Plaines. The northwest suburban perspective at WYEN helped me break perhaps the biggest story of my career. It also was the beginning of the end for me at WYEN.

On December 13, 1978, I received a call at WYEN from an officer I knew well at the Des Plaines Police Department, asking that I run a story about a missing fifteen-year-old boy, Robert Piest. At first I was inclined to take the information and do nothing with it, but the officer told me to go with it because it would soon become apparent that this was far more than a routine missing person's report or runaway. I prepared the story, which ran that afternoon.

WYEN continued to run the story, even though police gave us little inkling of what was going on behind the scenes as officers tailed contractor, party clown, and democratic precinct captain John Wayne Gacy and began to amass evidence through the execution of a search warrant at his Norwood Park Township home.

I'd continue to check each day with Des Plaines police. On December 22, my contact called again, saying that Gacy had been arrested on a marijuana possession charge, but that they expected something far more serious, something that may have a link "to the Piest case and a lot more."

At this point, the news began to break—fast. The morning of December 23, I had a conflict. WKQX needed me to work morning drive, as announcer and newsperson, because Program Director Bill Stedman was late for work; WDAI-FM had abruptly changed format overnight from album rock to disco, and Stedman had been on the phone for much of the night. WKQX was about to make a two-step format change of its own, from album rock to a home-grown classic rock, which I helped to program, on January 1, and to adult contemporary six weeks later. Stedman was trying to determine whether to pull the trigger on the WKQX format switch or retain album rock. Kaye filled in for me on morning drive at WYEN.

I checked with my Des Plaines police contact as an NBC engineer ran the control board and was told that Gacy was being charged with six counts of murder, a story that I promptly put on-the-air both on 'KQX and our sister station, WMAQ. I advised Beth what was happening. But he told me to call back once things began to settle down at NBC. Stedman showed up at 8 a.m. to finish his shift. I reverted to the newsroom and worked the story.

By the time I got back to my contact, he had information that was beyond belief. He said that Gacy had confessed to committing thirty three murders and had drawn a map that indicated most of the evidence would be found beneath his home. I advised Reiman, who declined to air it because my source refused to be identified. My next call was to Stew, who had the same reaction, even though my source had a close tie to WYEN.

I was off the clock at NBC, but was pursuing a career-making story. I spoke several times with my source, explaining the fix I was in. No one could believe the numbers I was prepared to report. Could someone, anyone, go on the record?

Hours passed. Other stations were still reporting that Gacy had been accused of killing six people, but WMAQ and WYEN both knew better, and I became more anxious by the hour that someone would steal the exclusive out from under me. Stew was the one to finally give the green light, realizing the close connection to WYEN that our source had and that he was completely trustworthy. WYEN had the scoop. Everyone else followed, and everyone wanted the story.

Associated Press AP and United Press International UPI called, both wanting reports. I made it clear that someone else from WYEN would have to voice the UPI reports. I fed AP Radio almost hourly at first as Bob Roberts, while filing reports for WYEN as Frank Rhodes. NBC Radio News made use of WMAQ's well-staffed local news department, and at first I fed NBC Radio on the story only if I was assigned to work that shift, but the well-drawn lines soon began to blur.

Dismantling Gacy's home and digging up the property took more than three weeks. I began my days during the dig by coming in to do the morning newscasts on WYEN. Then I went out to the Gacy home on Summerdale Avenue as soon as I was done with the 9 a.m. newscast, which dovetailed well with the investigators' schedule. It was an era before cell phones. I'd gather tape, retreat to my car in the increasingly bitter cold to write my reports, and then feed the reports and sound bites. Often it was through the graciousness of neighbors who became familiar with me, using a patch cord to connect my cassette recorder to their phones to transmit the sound. Sometimes, when I was really cold, I'd drive a mile or two to the Norridge home of my friend(and avid WYEN listener) Cindi Beck, where I'd get not only a place to feed my story, but something warm to eat and drink as well.

All seemed well at first, but soon the problems began to mount behind the scene. Everyone wanted my work. WYEN wanted reports all day. AP Radio Network wanted all I could send. So did WMAQ and NBC Radio.

I used "Bob Roberts" on both AP Radio and NBC, and network news directors in New York and Washington wanted to know how come I was on the competition's air, although I was careful to try to use different sound bites and write different reports.

Shortly before the dig ended, Reiman delivered an ultimatum from 30 Rockefeller Center. I had to make a decision—WMAQ or WYEN? NBC or AP Radio? Although NBC and WMAQ continued to offer only part-time work, they wanted an answer, and I had to make the decision that day.

I called Stew to let him know of my dilemma. I didn't want to make a decision. I liked working at both stations, valued the exposure AP Radio and NBC gave, liked the pay, and didn't want to give any of it up. I was still covering the digging, and as it continued around me, I continued to contemplate what I should do. Reluctantly, I called WYEN back, this time asking for Mr. Walters to tell him that I'd made a decision.

As I told him of my decision, he stopped me. He asked what it would take to keep me at WYEN. He offered a pay increase if I wanted to stay and even asked if I wanted to be news director. I told him no, for two reasons.

First, I respected Stew. I didn't want to see him stripped of the title or put out of work. It wouldn't be right. The other reason was financial. WMAQ could offer me more pay in two shifts a week than Walters offered full-time, even with a raise. I had network ambitions. The downtown, network-owned station had won. I told him that I valued my time at WYEN, but that it was time to go. As he said good-bye, he sounded hurt. I didn't feel good about the way I'd left it, but I didn't see a viable alternative. It was January 1979, and my time at WYEN had ended.

I stopped by a couple of times in the next few weeks while covering stories in the area, and Stew was gracious in allowing me to feed from the WYEN studios even though I no longer worked for the station. I left him sound bites (or as we know them in the business, "actualities") as a gesture of thanks.

The years passed. I moved to WINS/New York in 1984 and returned home to WMAQ in 1988. WYEN was sold. Stew changed stations, and I would *occasionally see him when covering stories in McHenry County. Wally would call me with stories from Payton's many enterprises before the Hall of Famer's untimely death. I addressed Beth's broadcasting students at Olivet Nazarene University's WONU-FM. Wayne Allen went into sales at WLS-TV, and I ran into him frequently when he worked as a cameraman during a strike against ABC. I'm a strong union guy, but I cut him a bit of slack I didn't show the other strikebreakers. Rob Reynolds became the ultimate pitch man with warmth and a voice I recognize anywhere. Spann bought WVON Radio and remained an R&B legend.*

One day, in 2006, I was assigned a story in the O'Hare Lake Office Complex that WYEN had called home. I entered 2400 W. Devon for the first time in twenty-five years. The building was under renovation. Wires were hanging down, the ceiling panels removed. Worn carpet was being replaced. The building looked tired. I hustled down the stairs, the same as the old days back in 1978, to the basement suite from which WYEN had broadcast. I looked through the glass door and saw dozens of boxes piled high. It was storage space. There was no sign on the door or inside that a radio station had ever been there. I closed my eyes and tried to visualize the WYEN I'd left in early 1979, the cramped newsroom, the people with whom I'd worked, and the resurrected collegiate air name that I have never since used professionally. I stood for a moment, then turned and left.

The old WYEN was gone. I have no airchecks of my work at the station. But Request Radio plays on, every now and then, in my head. It was a different era. As Mark Knopfler would later sing, "That's not work." Radio was fun.

Roberts became a full-time staff member at WMAQ in September, 1979, was laid off in 1984, and moved to New York to work for Westinghouse-owned 1010 WINS Radio as a news and sports reporter and anchor. Westinghouse bought WMAQ and brought Bob back to Chicago in 1988. Ten years later, Westinghouse purchased CBS, including competing all-news station WBBM Newsradio 780; it moved Bob to WBBM in 2000 when it closed WMAQ.

Chapter Ten
HE'S NO JOHN WAYNE

John Wayne Gacy.

Des Plaines police let Gacy's name saturate the large room they'd set aside for the media.

Before jotting his name, I'd been prepared for quite a story because of Bob Roberts work on the tips we'd received. Only when I saw the large number of police brass standing behind the podium did I grasp these tips were not exaggerated and would now be presented as facts.

Police divulged enough in their first news conference on the *Killer Clown* to make this room of reporters recognize Gacy news demanded constant updating. Details were very fluid of his arrest, criminal background, and life. As police dug up victims from the crawl space of his home, the media also recognized how fluid the information was on the list of young boys he killed. Over a period of time, we'd learn police had tailed Gacy, day and night, after they identified him as a prime suspect in the disappearance of fifteen-year-old Robert Piest of Mount Prospect, Illinois. Once released, tidal waves of all things Gacy splashed everywhere. His name and the multiple charges against him flowed outward in waves washing over Des Plaines, flooding Chicagoland, and finally unleashing an unrelenting torrent on every spot in the world where people watched television, listened to radio, or read a newspaper.

In the middle of all this seriousness at the Des Plaines Municipal Center, I wondered whether I'd pressed the record button on my recorder. Television and radio news people stuck their microphones on the podium, some duct taping to the main microphone for police. I taped mine to the podium after I saw the main microphone nearly disappear in a dozen Electro-Voice microphones taped around it, creating what looked like a metal football. Normally at a news conference, the radio reporters tend to sit a few feet from the podium so they can get up quickly and rush to push the record button just before the speaker begins talking. But at the Des Plaines Municipal Center, the reporters sat at tables. Des Plaines city leaders likely had meetings in this room where you got a lecturer or a city supervisor, and everyone else sat down with their notes at long tables and within reach of water pitchers.

One or two television cameras between the podium and reporters made it possible for me to slink by, but a phalanx of television cameras set up within minutes, a virtual wall of cameras, stopped even the most daring reporter. You just did not walk in front of the cameras, and this rivaling the Wall of China in television cameras prohibited any radio recorder check beyond the initial setup.

My focus shifted from the record button. As the news conference pushed on, I got stuck mentally on the name of this accused serial killer—John Wayne Gacy.

"John Wayne ... John Wayne ... John Wayne Gacy," I whispered his name. Maybe saying John Wayne Gacy wouldn't bother me so much if I kept repeating it, but I couldn't shake thinking of the great movie actor John Wayne. I'd been a fan of the Duke long before he won an Oscar in 1970 for *True Grit* or performed in his last movie, *The Shootist* in 1976.

Perhaps I could leave out Gacy's middle name in my stories? I actually wrote his name in my notes as John Gacy. I suppose it sounds kind of trivial to read about this now ... so many years later to let this be such a big deal of the name thing, but back then, the newness of the Gacy arrest played with our sensitivities, and I was in need of separation between John Wayne and John Wayne Gacy. But even more than the John Wayne name problem I had, something more intense was happening. The sheer number of Gacy stories helped hasten in me something fairly common among announcers that read story after story of death. We were slowly desensitizing, unintentionally removing ourselves from the emotion of death, and in some ways these stories were just that ... stories ... and not of people who somehow touched our lives. But at that moment, with the John Wayne dilemma of how to say Gacy's name, I still was truly bothered. I'd battle desensitization for much of my career.

Some media outlets printed the sickest Gacy details. However, I'm convinced most journalists did not delve into detailing Gacy's life and crimes in a clinical way. Though tempted by larger readership or more viewers and listeners, I believe the majority of the media drew a line of decency they would not cross.

The backyard of WYEN's signal swallowed Gacy's West Summerdale Avenue home. WYEN wasn't too far from Norwood Park, where Gacy served as a Democratic Party precinct captain.

Gacy's sick saga brought the full attention of WYEN. Gacy preyed on young men and boys in the neighborhoods where WYEN's signal was strongest. We delivered Gacy stories, an obligation I pledged to our listeners. Not only were we standing in front of the Norwood Park Township home with updates on police excavations on the Gacy property, but we were first in breaking the depth of this story, thanks to the source close to WYEN and the efforts of Bob Roberts.

In preparing for *The WYEN Experience*, I asked the source of the Gacy information to let me reveal his identity because so much time had elapsed. I thought why not, it shouldn't matter any longer. However, the response completely surprised me. The person could no longer remember any of our conversation. I will just have to tell you how the conversation played out.

I was sitting in the WYEN newsroom typing stories. Someone from outside the newsroom called my name. I headed for the front of our office where I found my source.

"Stew, you or one of your news people should be at the news conference tomorrow at the Des Plaines police station. I believe it's for 10 a.m."

"Sure, I can do that."

I appreciated my source letting me know about a news conference. This departed from the usual means of finding out about a scheduled event. The media generally relied on City News for its assignment schedule. This time I received a personal invitation, so I attached importance to whatever it was I would be covering the next day in Des Plaines.

"Is there anything you can tell me?"

"Des Plaines police made an arrest. I can't tell you more about this other than it's a murder investigation."

This really struck hard for what was not said. I had never received such a tip before. Usually the mention of a news conference at a police department on the City News Wire followed a news release on the arrest of someone for murder. This time, I never saw a news release, just a source saying, "*Show up at the Des Plaines Police Department tomorrow.*"

"This is a murder investigation you said?"

"Yes."

How far could my source go? I'd push a bit.

"Was there more than one murder?"

I couldn't imagine getting called out of the newsroom and seeing my source for this information and not having more details. I know this sounds crass, but I threw out numbers.

"Two, three, four, eight, nine, ten?"

This wasn't my usual way of gaining information. I was uncomfortable.

"No, there were more," he answered, but stopped abruptly.

I didn't think he'd answer, but he did.

This internal process of silently going back and forth is getting in my way. He's not leaving! This is big. Ask again.

"Twenty murders?"

My source didn't budge.

"Twenty-five murders?"

Again, no movement.

"Thirty murders!"

Finally, my source reminded me of the news conference in Des Plaines. He certainly did nothing beyond what any other officer might do if asked about a news conference. I wasn't sure whether I had a huge story or a pretty big one. I narrated an unusable story. My source hadn't left, giving me one more chance for his reaction.

"Multiple murders in the suburbs! Police in Des Plaines have arrested a suspect and will name him tomorrow at a news conference in the north suburb. I'm Stew Cohen, WYEN Metro News."

I didn't get a reaction—definitely not a yes or a no. I couldn't use the story because it wasn't supported. The additional information Roberts gathered helped change my mind.

Over the past thirty-two-years, I've kept some of the Gacy copy I typed on half sheets of yellow paper.

> ***Dated January 5, 1979****: Cook County Sheriff's investigators and county highway workers continue the tough job of digging in the crawl space of John Wayne Gacy's home. So far, twenty-seven bodies have been found under and around Gacy's Norwood Park Township home. The thirty-six-year-old contractor will be in court next week for a preliminary hearing, but already, a county grand jury may begin considering evidence in the mass murder case.*

> ***Dated February 7, 1979****: Investigators are going to court to try to get permission to make public personal items found in the home of accused mass murderer John Gacy. The effort is intended to help identify more of the bodies excavated from Gacy's Norwood Park Township Home. Police say they are disappointed that so few people have sent in dental and medical records of missing persons to aid in identifications.*

Roberts described in chapter 9 how he stood outside the Gacy home in the coldest winter in years. It got so cold, police occasionally postponed digging outside the home because the ground was frozen. The only time Bob missed reporting from the street near Gacy's home, I agreed to file stories.

The surreal nature of Summerdale Avenue struck me even before I got out of my car. I parked down the block from the Gacy home. From inside my car, I looked at everything within view. The windshield of my orange Vega hadn't frozen yet. People stood on the sidewalk opposite the house. They wore so many layers of clothes they looked like coat racks holding two or three coats piled on top of each other. The absolutely brutal temperatures didn't seem to make a difference to these people because their concentration on the goings-on within the Gacy property was intense. While people may have appeared oblivious to the arctic chill, I regretted getting out of my car. My knit hat, scarf, thick gloves, and heavy coat gave about thirty seconds' worth of protection. Thirty seconds wasn't enough for doing justice to the story. It needed ninety seconds, but I could only give sixty seconds and still have enough time left for a handful of other stories. The idea of writing notes held merit. Who'd write the notes? I wasn't taking my gloves off for a second.

Bob Petty of WGN TV came from out of nowhere it seemed. Funny how I remember more about Petty's clothes than I do about the actual excavation work by police at the Gacy home. Petty stood right in front of me. I had my own personal TV news set in this frozen tundra with the backdrop of Gacy's home. Petty made an impression. He was the better man for withstanding dangerous cold while dressed GQ style. He wore a long camel coat with leather gloves and an expensive scarf wrapped around his neck … but no hat. My ears tingled in my knit hat.

Just endure the arctic deepfreeze a few more minutes, I thought, enough time for his full report and to watch Petty's handoff back to the studio. The WGN TV camera crew moved into position, switched on the lights, and Petty brought the microphone to his mouth.

"In front of the Gacy home on Summerdale Avenue, there is little activity now, while we understand officers are digging in the crawl space, looking for more bodies." His exact words were impossible to remember simply because my ability to function was breaking down.

Cold air billowed from Petty's mouth; he was a human smokestack. Clouds pushed out beyond his microphone and vanished. I watched, listened, and waited for something he reported that revealed new information. Petty finished, the lights turned off, and for me the show was over. I thought back to what he said, tried to analyze and see what might help my story, but all I knew were two things. I wish I'd worn long underwear and brought a ski mask.

I went back to my car and sat for a few minutes with the car running and the heat blasting on my hands. When enough feeling returned, I gripped the steering wheel, drove to the radio station, and filed wire service stories on Gacy. (Note how the network spelled my name wrong.)

Audio-Hourly Newsfeed
Following Cuts to Be Fed at 2:10 and 4:10PM EST
48:35 V DESPLAINES–(Bodies)—(Stu Cohen) 3 Bodies of Boys Discovered, Police Search House For More.

In my research for *The WYEN Experience*, I learned that a member of the WYEN family came in contact with Gacy. This chance meeting occurred some years before he began working at Request Radio. As I understand the story, a drug store in Morton Grove needed renovation, and Gacy came out as a contractor to see what he could do. He got around to talking to one of the employees and invited him to dinner. The answer to the invite was a resounding no! In my interview, the WYEN staff member said he thought Gacy was creepy. Later, he learned what Gacy had done. This is eerily similar to Gacy meeting Piest.

Gacy's trial didn't start until after I had left WYEN and moved to Crystal Lake, Illinois, about an hour north of Chicago, and began working for Mal Bellairs at WIVS-AM and WXRD-FM. A change of locations didn't make the Gacy story any less important, but for some reason, I never thought to pick up the phone and try to schedule an interview with Gacy's lawyer, Sam Amirante. However, thirty-one years later, Amirante and Danny Broderick cowrote a book, *John Wayne Gacy, Defending a Monster* by Skyhorse Publishing. In the years between defending Gacy and today, Amirante became a Cook County Court associate judge and then retired and returned to private practice with offices in Palatine, Waukegan, and Chicago. Broderick has handled felony and misdemeanor crimes at the Law Offices of Daniel J. Broderick. Amirante and Broderick could now answer Gacy related questions.

"Judge, what were your impressions of John Wayne Gacy?"

As far as him as a person, you don't have to like your client. John Gacy was a liar, a manipulator, a con artist, a murderer, but he was my client. There's a certain bond a lawyer has with his client—a professional bond. You have a duty to the client. People ask if I went to Gacy's execution. I did not. I did not want to see my client executed. Did I like him? Initially when I met him, I didn't dislike John, but I can't say I liked him. I had no feelings for him either way. He was somebody everyone knew. Everybody thought he was a good guy in the community. After I began to represent him and after I heard all the things he did, did I like him? Absolutely not! But there's no requirement in the law, ethics, or anywhere else, you have to like your client. You still have to do your best. That's the thing about being a lawyer; you have to do your best.

"Judge, you were despised for representing a monster. Many people thought Gacy didn't deserve a trial; just send him off to prison and the chair."

He had a constitutional right to a fair defense.

(In their book, Amirante and Broderick wrote in detail on the American Justice System and how some people were so upset with Amirante for defending Gacy, they threatened him. He said those same people would be first in line for an attorney if they were in trouble.)

"Were you distracted by the media circus outside the Gacy home?"

The interesting thing, and there were so many media personnel, so many police officers, so much going on, and I don't know if it's something I was educated to do as an attorney, or just my personality, I sort of focus on what's going on in front of me, sort of like a surgeon, just looking at the spot of surgery, not paying attention to crowds or media. I was focused on just hoping my client could get a fair trial later on and not have a circus atmosphere or this much media attention to it. But I was focused on the evidence and what was going on in the case, rather than worried about the media. I really didn't notice it.

"Judge, I understand one good thing did come out of the whole Gacy case with response time on missing children."

It's not that the police did not care. They had priorities—murders, armed robberies, rapes, auto accidents. Most missing kids have run away, especially teenagers because they had a fight with their boyfriend or girlfriend or because they are upset with their parents or there's something else going on, so police thought these kids were going to come home anyway. This wasn't a top priority for them. They just didn't think of the kids that fell through the cracks, the kids that didn't come home or wound up in the hands of someone like Gacy. At the time of Gacy, there was a whole different attitude toward homosexuals in the country, in the world. Somebody would go into the police station that wasn't very macho and say, 'This guy attacked me.' Unless it was a child saying this, the police would generally say, 'You'll take care of yourself,' and brush him off. What I did with the I-SEARCH Law in 1984 was try to get rid of an attitude the police had that all kids are just runaways and are coming back. There might be something wrong, and the most important time in looking for that child or young adult is in the first seventy-two hours, and if they'd let it go, the kids might be gone or missing forever. If it was my kid, I'd want the police out there busting their hump looking for him. That's really what happened in the Gacy case with the Piest family. They went to the police department. They said they had a good kid. They wanted the police to work on it right away. They dogged the police ... they wouldn't give up. Unfortunately, even the seventy-two hours didn't matter, but what I found in my experience was the initial work that the police did was so important in finding these kids and bringing them home and maybe finding somebody out there that might kidnap them. A lot of the law later went to parents who have kidnapped kids, and it went into a different area because parents could leave the state, and it would be difficult to bring the kids back. A uniform child custody act came into play. The parental kidnapping act came into play, and a lot of good things developed out of it like the Amber Alert law.

"If all of these things had existed back in Gacy's day, Judge, do you think it would have been more difficult for him to do what he did?"

It would have been more difficult because of greater awareness. Parents went to the police, and they were ignored. Complaints went in about Gacy that no one knew about. His neighbors didn't know about the complaints. Parents of missing kids would have to hire private investigators to try and do things on their own, and they'd spend a lot of money.

"Danny, you recall a case?"

The family of John Sync hired a private investigator for $5,000. The Sync family felt they had to do this rather than go to the police. The police did not treat it as a high priority.

"Judge, you wanted to say something?"

To answer your question, Stew, the laws since Gacy would have made a big difference back then. A lot of those kids might have been saved had there been more pressure on the police departments. What we basically did at the time was develop a law that did three things. It ended the seventy-two-hour waiting period, started a statewide central computer system, and formed these I-SEARCH units. With the computer system, all information on lost, missing, and runaway children goes into the computer, and all information we know about sex offenders and kidnappers. Now there's a whole registration process. I-SEARCH units out of local police departments specialize in finding lost, missing, or runaway children and make the community aware. This was the result of me representing Gacy.

I did not press Amirante on the details of the murders or focus on that aspect of the story during my time with WYEN. I hadn't intended to change my approach in my interview with Amirante and Broderick. Instead, I recommend reading their book, *John Wayne Gacy, Defending a Monster* for an incredible read on Gacy's thinking and methods, and how Amirante worked on his defense.

Shortly after the state put Gacy to death by lethal injection, Mal Bellairs, my friend and radio mentor from WBBM AM and later owner of the McHenry County radio stations in Crystal Lake and Woodstock, told his radio audience he attended one day of the trial at the invitation of a member of the prosecution. William Kunkle gave Bellairs some wood splinters from the trap door used by Gacy to move his victims under his kitchen into the crawl space. As Kunkle's guests, Mal and his wife, Maria, were permitted by Judge Louis Garippo to sit inside the area where defense and prosecutors worked in the large courtrooms at Twenthy-Sixth Street. They sat very close to Gacy. Mal called it an eerie feeling.

Over the years of working in radio, I've covered many murder cases. Some of the cases were big locally, but nothing approached the impact of the Gacy case. That news is a relief.

Chapter Eleven
THE BIG BOYS

Garry Meier

A leg dangled from the air studio's ceiling. This was not the leg lamp kind of leg from *A Christmas Story*, but more of a stripped-down female leg that may have been removed from a mannequin. Mr. Walters gave me a tour of the on-air studio on an August day in 1976. The only thing I can clearly remember from that long-ago radio station tour was the dangling leg. I wondered what type of personality might be behind its placement in the ceiling just to the right of the announcer. I mean, my initial thought was that my ability to concentrate was less than fully developed, and now I'd sit in this studio, if Walters hired me, and try to read my newscasts, and in the corner of my eye I'd see a dangling leg sticking out of the ceiling. An off-kilter sense of humor was behind this, and although some believed Garry was responsible, I'd never know for sure until my interview with him for this book.

Thirty-five years after "Leg Gate," I sat in Meier's office at WGN AM 720 in the Tribune Tower in downtown Chicago for a one-on-one interview with the former WYEN midmorning announcer. I hadn't brought up the leg until the end of our half-hour interview. I asked him to talk about the type of equipment used in the WYEN studio, and he talked about using a cassette machine instead of a four-track cart machine. But he also discussed the frustration some of us had because we didn't have a single window looking to the outside.

"I remember that studio, kind of drab and no outside window, so I decorated the room with a bunch of posters of sunsets that someone gave me. The leg came from the drug store where I had once worked, and I had it coming out of the ceiling tile in the corner so when people walked into the studio, they thought someone was coming out of the ceiling."

This "leg in the ceiling" thinking was just like a window into Garry's personality. He knew his personality was as he described it, "a little different than most people around me." Not that Meier needed someone vouching for his God-given talent, but a lot of us at WYEN recognized he had the type of personality and brain wired to visualize things in a more unusual and often humorous way. We heard moments of brilliance on-air, but some of his talents were shown off-air in a weekly newsletter he created under the name of Tommy Turntables.

"This was a sarcastic run of the station where I wrote about stuff happening there and just twisted it in the way I do today, and I'd put it on the bulletin board inside the studio."

CBS sitcom *Dennis the Menace* and its child star Jay North and legendary WLS announcer Larry Lujack were Garry's "tipping points" as he described the key moments that led him to a life in broadcasting.

"In the late 1960s, Lujack would read a letter from a listener every day, and he'd always talk about reincarnation. He said he'd come back as a butterfly. One day, I found a big Monarch butterfly on my lawn, dead, and I sent it to him and said, *Sorry, I think I killed one of your relatives.*"

Lujack read Garry's letter on-air, and when he heard it on WLS, Garry knew he had to get into something like this because there was his material being used on-the-air.

Jay North was Garry's age and living the kind of life Meier soon discovered he wouldn't mind. The show *Here's Hollywood* captivated a very young Garry, especially the episode where the cameras were brought into the North family home and property and captured North in his inground pool sitting in a little rowboat going around and around. Garry interpreted his memory as a child, "Somebody's getting paid to do that, and they get to live like this!"

Had his career track shifted slightly, Meier might have been the guy in the pharmacy dispensing medicine to customers. He was in pharmacy school, but in his mind he still thought about radio. He had no idea how to get into radio, and as he told me candidly in our interview, "I was just trying to stay out of Vietnam basically, and I thought I could get a deferment by going to pharmacy school. I got a high lottery number, and that's what kept me out of the war because I dropped out of pharmacy school." But instead of sliding into radio, he worked construction for a couple of years because he was "tired of being poor." Construction, though, wasn't his future. He just wasn't good at it. Garry recognized he lacked the skill set.

"I looked in the *Yellow Pages*, and I found broadcasting schools, and I started to go to one, where I met a guy who got me a job, and from there, it all started to snowball."

WFYR-FM in Chicago offered Garry his first job, and he ended up doing the all-night show. "But it was automated," Garry remembered. "You just changed the tapes and sat there and babysat tapes, and I thought this was not why I got into radio."

He heard after his first year at WFYR some rumblings that a station in Des Plaines was hiring. "It's live where you do everything, and I thought, *That's what I want*, so I sent a tape to WYEN. I followed with a call and was brought in for an interview and got hired."

So began Garry's radio career. For the next three years, 1974 to 1977, he worked as an announcer on a live radio station.

While some former announcers looked at WYEN as a training ground, Garry was not one of them.

"We tried to beat the Chicago stations. I don't start anything with low expectations. I went to WYEN like I was going to the biggest station in New York in my head."

Meier recognized WYEN was a great foundation for what followed in his career.

I asked Garry for a memorable moment at WYEN. In typical Meier style, he launched right into a memory that was descriptive, funny, and embarrassing.

"I was breaking up with my girlfriend after this long-term relationship, and one night I got tickets to a movie premier downtown, and I decided to call her. I thought maybe we could massage our relationship back. We went to this movie, and afterward, we got together with several people from WYEN. I got very drunk, and she had to drive home, and I ended up sleeping on a couch at an apartment she shared with her cousin. I took out my contact lenses and put them on a table wrapped in some tissue. The next morning, her cousin started cleaning up the place and threw my contacts away in the trash! As hung over as I was that morning, I searched in the trash for my lenses because I had to have them to drive to work. Well, I only found one. I washed it off and put it in. I started heading down the Tri-State Tollway, but the drive became an adventure as I started feeling sick, basically throwing up in the toll baskets—and I'm not exaggerating. I finally made it to work, but I was a mess. My clothes were all screwed up from vomit. Someone told me I'm on in an hour. I thought I could do it. Walters gave me a pair of his pants, though he was about three inches shorter. About an hour into the show, I could barely talk, and I think it was fairly obvious ... so Program Director Wayne Allen came into the studio and told me to lie down in the conference room. What could I do? I went in there and stacked probably fifteen albums for a pillow on the floor. I lied down for what I thought was an hour, but was actually four or five hours. There wasn't anyone around when I finally woke up. This was during the Christmas season of 1976, and I found a bonus check on my chest that Ed must have dropped on me, and I still had his pants on. I think that was rock bottom. In hindsight, this was one of the funniest things I can tell about WYEN."

Money certainly had a place in the conversation of the announcers I talked to for *The WYEN Experience*. Garry was not the exception here, however, he didn't dwell on salary because Rob helped to bring him into focus on his career goal.

"WYEN was a suburban radio station, and I don't know what we were thinking, but I'd get all heady about the need to make more money. I was earning one hundred dollars a week, and although it was the 1970s, one hundred dollars a week wasn't good money even then. I think I expected too much out of WYEN."

Reynolds had Meier's attention. Garry respected the fact Rob was a Northwestern University graduate and felt his advice kept him centered.

He recalls Rob saying, "WYEN is not going to be a great paying place, but it's not going to be the end of your career, so take it for what it is." Garry knew this was the perspective he needed, and things were fine then.

Garry, Wayne, and Rob wouldn't stand in the way of making more money, especially with an opportunity to do what they did best—play music. Each benefited from working a second job at the Savoy in Wheeling, the old Union Hotel. Garry was extremely relieved that Rob got him a job spinning records at the Savoy for ten dollars an hour. He knew his WYEN salary didn't go very far. Garry had to go through so many tolls living in south suburban Oak Forest that things were very tight.

"All of a sudden, I went from one hundred dollars to four hundred dollars a week, and now I'm almost intoxicated by that, and I don't want to leave."

In November 1974, Garry stepped into WYEN studios for the first time and believed he'd be there a year and then move to a station downtown. He stayed three years and only left because he was pushed out in September 1977.

"I think I was getting frustrated and taking it out on people and management at WYEN. I should have just shut my mouth as Rob had advised. Wayne called me in one Sunday, and I got wacked. I was already prepared. I had cleared my desk out, a process some people do to try and move things forward. I cleaned it out a month before they fired me, so when it happened, it was okay. I said thank you, and I was able to go to my car without stopping anywhere along the way. I recall some people were surprised how jovial I was when I got fired. They thought I'd be a little more depressed, but I knew it was time for me to go, and I couldn't make the move on my own; I just needed to be pushed."

Though Garry could not leave WYEN without management's decision to make a change, he was thinking about his future every day.

"I'd been talking and sending tapes to WLUP-FM, The Loop every other month. I'd send a tape or make a phone call, and they finally called me and said, 'We have a job, a public affairs job on weekends, basically doing interviews with community leaders.'" Garry didn't accept the job. This was not what he was looking to do, but it was all WLUP had open.

"I casually mentioned this to Chris Devine, and he asked if I could recommend him. I did, and Chris took the job. The day I got fired, I called Chris and said, 'Hey, is there any chance you can talk to the program director at The Loop and see if there is anything?' The next day, I got a call from the program director, and he told me to come on down. They hired me!" Garry started on weekends, and then it became an all-night job he said "that set the course that I am still on."

Garry went from WYEN to WLUP-FM, which he says wasn't quite The Loop that people remember with Steve and Garry and Kevin Matthews. The time would come. The station, as Meier remembers, was hip and was a music station. He went from one hundred dollars a week to 250 dollars a week.

"I'm thinking, *This is it*, and I'm happy as hell to be there!"

Meier started at the Chicago radio station in the fall of 1977 on weekends and slipped into all-night and full-time. Within a few months, there were rumors the Chess family was planning to sell the station. (Garry says that in the early to mid-1970s, Terry Chess was general manager of WSDM. He is also the son of Phil Chess, president of the Chess, Checker, and Cadet Records' radio company.) Garry recalls how the rumors of an impending sale of The Loop led to panic, so they called a meeting, and Chess said he wouldn't sell WLUP unless someone offered him five million dollars cash, which Garry notes was pretty big money in the 1970s for a radio station. About a month later, a memo came out.

"We just sold! Someone offered five million dollars cash. Former Hawaii Congressman Cecil Heftel bought it."

As Garry tells the story, he remembers, "The place panicked because everyone thought Heftel would change the format. People thought he'd fire everybody."

Panic overcame Garry initially, but he quickly calmed down and realized they wouldn't be firing the all-night guy; he was the last guy they'd look at because he had no bearing on anything major.

"This was an unpleasant time because everybody thought they were going to get fired." Garry went about doing his job, enjoying the all-night show. The new owners brought in Lee Abrams, a well-known consultant.

"Lee decided to bring Steve Dahl over to do mornings. I did all-nights. Steve and I met on-the-air, and the rest they say is history."

Garry still has fond memories of driving by the O'Hare Lake Office Plaza off the Tri-State where the WYEN studio was and thinking of what happened downstairs in the basement of the building at that studio.

Before ending my session with Garry in his office at WGN AM 720, we got serious for a moment. Like any broadcasters that care about their business and craft, we can discuss the industry as we see it unfolding before us. I wanted to know what he believes is happening to radio.

"The challenge for people who want to get into the business now is to be willing to deal with the low pay and struggle that it usually takes to get to a big market, if that's your goal. I don't know how many WYENs are out there anymore where people can smooth out their rough edges, and some people might not want to put up with the crap a lot of us did thirty years ago, and that was before all the new technology that is part of the swirl now."

Final thoughts from Garry to end our session on WYEN:

"If you want to get into anything, you find the WYEN of that thing and you go and enjoy every minute of it. You think you want to get out of it like you may have felt about high school, until you get out, and, boy, you're now in the real world, and your head may get stepped on ten thousand times. You think back and remember those enjoyable moments when things were so much less pressurized. I look back and enjoyed the whole run of it. We had a great time because we were all on the climb up the mountain. When you get to the top of the mountain, then you are in for your real battles. I didn't know that until I'd been gone from WYEN for many years."

Meier and Dahl

Sears and J.C. Penney are anchor tenants of Golf Mill, an open-air shopping center at Golf Road and Milwaukee Avenue in Niles that's stood since *The Sound of Music* played on the huge single screen at Golf Mill Theater in the mid-1960s.

Golf Mill and its shoppers represented everything that embodied WYEN in the 1970s. The shopping center competed for attention with Randhurst and Old Orchard in the north suburbs, and this was before strip malls became popular, affecting the survival of the larger indoor malls. I can only recall one Golf Mill Radio Days event in the 1970s, probably because I hadn't paid attention to this type of event. Mr. Walters had called for every full-time staff member to join him at our table. The radio stations throughout Chicagoland set up tables from just outside J.C. Penney on one side to Sears on the other. Request Radio was positioned near WLS-AM. John Records Landecker manned the WLS-AM table.

WYEN gave out lots of helium balloons and bumper stickers, but we were a little picky with our WYEN T-shirts. Occasionally one of the announcers sucked in a bit of helium and sounded like Donald Duck. Our music played at our table, and so did music from a dozen other stations, all mixing in the air, creating a potpourri of sounds that blended into noise pollution. After a while, Meier, Diamond, and Val Stouffer decided to walk around Golf Mill and check out the other radio stations. I joined them. Walters and a few others stayed behind at the table. He didn't seem to mind us leaving our table. The four of us started walking toward the Mill at Golf Mill. We passed several tables where radio stations had their tables and gave out bumper stickers, balloons, pens, and shirts. We saw WBBM, WGN, and other well-known stations, but we didn't stop. Something drew us further. Maybe we just wanted to know the extent of Radio Days at Golf Mill, and so we continued on toward the Mill. I was more a follower, talking with Dan while Garry and Val led our little group. Beyond the last of the radio tables, we approached a young man, flanked by two beautiful women. No music in the air at the Mill, just these tall, blonde, exceptionally out of place women that I could not stop staring at until the young man started talking. He was casually dressed, husky, and friendly. He reached into a bag and pulled out cards with his name.

"This is my Breakfast Club card," he said, smiling at Garry, Dan, and Val. I was staring at the card, trying to read it before I stuck the card in my pocket.

"Have you seen all the tables?" he asked. We said no, but we'd seen a few. None of us asked his name, but we didn't have to because the card had his name printed on it … Steve Dahl.

Greg Brown

He didn't stick around for very long at WYEN, but Brown, an announcer at 94.7 WLS-FM, Chicago's True Oldies, weekdays from 3 p.m. to 7 p.m., has made a name for himself in Chicago radio. I asked Greg to remember January of 1972.

I really only worked at WYEN for five shifts. I was going to Millikin University in Decatur at the time, and I was going to have the month of January off, so I thought I would try to line up a job in radio for the month to get some experience and develop some contacts as well. I had already worked at WSTK in Woodstock, WVFV in Dundee, and had helped to build the college radio station at Millikin, WJMU ... so I was ready for the big time in my mind.

I called Ray Smithers and set up an interview at WYEN. As I traveled to the station for my visit, I had to drive partway on the expressway, a sure sign that this was the big time! When I arrived at the building, it was a huge office building ... another good sign. And when I walked into the station, it had a waiting area. Things kept getting better. Ray took me on a brief tour, and I couldn't believe how nice the studio was compared to every place I had been before. My hopes were high!

I told Ray I was only available for the month of January, which didn't seem to bother him at all. He told me I would start the next weekend on Sunday night, and then I would do Saturday from 6 p.m. to midnight and Sunday from 6 p.m. to midnight for the following two weekends.

The first time I opened the microphone and heard myself in stereo ... I was blown away! "In Stereo" was the big slogan as it was a big deal to broadcast in stereo in those days! My voice sounded so rich. The music was so full! My dreams were coming alive!

I took requests from people who were listening from all around Chicagoland! I had never been on a station that had a signal big enough that my family could listen to before, and I was thrilled to be able to play a song for my cousin ... who was as excited as I was!

Ray recommended me to Art Roberts at WGLD in Chicago. Art hired me, and we developed a great friendship over the years and later worked together at Q101 in Chicago.

I've been blessed to have worked my whole career in Chicagoland. Those five shifts at WYEN gave me the confidence that I could pursue a career in radio.

CHAPTER TWELVE

LESSONS LEARNED

The Body in the Backseat

"*Saeiho ahio onf aklsoineyi,*" the police officer spoke in plain English, though the words sounded like he was talking in a foreign language. The officer told how the woman's body was found under a blanket in the backseat of a car. I nodded, but I wasn't listening well enough to gather the facts. I only knew these things about the crime because I taped everything and later listened back on my recorder.

I parked my car in a spot near the crime scene at Golf Mill Shopping Center. Dozens of people had formed a tight circle around an officer who guarded a compact car as evidence in a murder. This was 1976, the end of summer on a day that started out cool and quickly warmed. I picked up my recorder bag on the passenger seat. I lifted the tan bag out of the car, put the strap around my shoulder, unzipped the bag, and connected the microphone to the recorder. Couldn't go anywhere without the bag that carried the *heavy as a brick* Marantz recorder and microphone, cassettes, and D batteries. I took my leather jacket off but grabbed an extra pen for notes and dropped it in my bag.

"Officer, explain what we have here."

"Are you treating this as a homicide?"

"Any type of weapon found in the car?"

"Did you find blood in the car?"

I wasn't asking anyone a question, just practicing what I might ask. The questions all sounded workable for a pool of questions I could dip into if my brain froze.

"Excuse me, excuse me," I politely said, gently stepping closer to the police officer. People were moving out of my way slowly, sometimes reluctantly, but they were moving, and then I faced the police tape. Actually, I didn't have to watch most of the episodes of *The Streets of San Francisco* to know you don't cross the yellow police tape unless given specific permission. I moved around the outside perimeter of the police tape until I faced the officer. He looked at me. I couldn't quite read the name tag on his uniform.

"Officer, I called the police department. They told me to talk to you, here. I'm Stew Cohen of WYEN Metro News."

The officer wore reflective sunglasses, probably part of the police uniform. I could see my eyes in his glasses, but I couldn't see his eyes, so I didn't know if he was a willing participant or annoyed that a pesky reporter had circled the police tape before buzzing under his chin. I mentally checked off the tools of his uniform. He carried a police radio, packed a service revolver and baton, and wore a badge. He was fully loaded and coming toward me. That was my cue not to waste his time.

"I have a few questions."

"Sure, come through."

"I've got my recorder here," pointing at the bag. "Just take a minute. Thanks."

I reached in and pushed the record button and pulled out the microphone.

"What can you tell me about what was found in the car?"

He leaned in slightly, and I stretched my arm further, gripping the long neck of a good Electro Voice field microphone, pointing the built-in wind screen toward his mouth like I'd prepared for a fencing match. For a few moments before he leaned in, he had blocked the sun on this bright blue day. But now, the rays hit me in the eyes, and all I could do was squint. I tried an advance lunge with the microphone. He didn't move his head back. I nearly hit him in the mouth.

I stressed on the effects of the sun. My eyes watered, my sinuses acted up, and my concentration wavered.

A woman stood leaning into the police tape, trying to hear what the officer might say in my interview. Another woman held a camera in her hand, intent on snapping pictures. I looked at these women and others behind the line and missed the officer answering my first question.

"Can you please explain a little further?"

"Hripdn rtqbf onnmri," the officer added.

"Where do you go from here with the investigation?"

"Niles Police will send the body *auoho wuye paeinwi*, and wait *weuiouho xsiu qiond*."

"What do you do with the car?"

"We'll treat it as evidence and tow it to the police station."

I politely threw out questions as though I were throwing dice, hoping not to roll snake eyes with crap questions he'd already answered.

"Is this the first murder here?"

The officer went on and answered … and answered a couple more very general questions … and I heard most of those answers. I just hoped my recorder worked and could play back the initial answers I didn't really hear.

"Thank you for the interview," I offered as my way of trying to add a modicum of professionalism on a dismal first crime interview.

The radio squawked. He just nodded. I heard a street name and number, and he then brought a hand up to the radio and cupped it toward his mouth and began speaking. Nothing I could hear, and anyway the interview was clearly over. I turned around, bent under the police tape, made my way through the crowd, and walked back to my car.

My Vega was my office on the road. I put the Marantz on the passenger seat, pressed rewind, and opened my notebook. The soft whirring of rewind stopped, and the button clicked and sprang back up. I pressed play, turned up the volume, and the officer's first words spilled out of the small speaker in the recorder. I began writing. His short first answer begged follow-up.

"A man walking to his car in the Golf Mill parking lot called Niles Police. He reported what he thought might be a hand slightly exposed under a blanket bunched up in a heap in a backseat. We found the car, cordoned off the area, and had the medical examiner remove the body and blanket."

Then I heard my question, "Can you explain a little further?" I'd finally find out what he said. This was no way to do an interview!

"We believe the woman was in her early to midthirties. She may have been killed somewhere else before her body was put in the car. Our investigators did not find much blood, though we believe the suspect used a weapon. Witness statements indicated the suspect ran though the parking lot and may have boarded a bus. The body is at the Cook County Morgue awaiting an autopsy."

I felt relief. Enough here for a story, and certainly Wilkinson didn't need to know I only got this information because my recorder was working. I didn't want to think what I'd have to do if I didn't have the interview on cassette. I vowed to improve my listening skills by doing whatever necessary.

My second major crime story had Bill directing me to a suburban office building where several men were shot to death in the building's parking garage. Again, 1976. Maybe the horrendous nature of this sparked my interest, but I heard every bit of what was said by police from the moment the officer started talking to his asking, "Any more questions?"

From all the trouble with listening and concentration skills early in my career, I learned what not to do and developed a lecture for students aimed at improving their listening ability. Spending time in community service was something Mr. Walters appreciated from his staff, though he didn't demand any of us lend our skills in volunteer work. He was just glad we gave back to the community.

I'm sensitive to the difficulty students have with listening, and so I offer an interactive lecture to teachers interested in giving their students insight from someone outside education. One of my favorite places to pass on the art of listening is on my yearly visit to Michelle Kilb's classroom of sixth-grade students at Lakewood Elementary School in Carpentersville, Illinois. My time in Michelle's classroom isn't scripted; I just wing it, hoping to cover important points of listening and writing skills. I also discuss the benefits of higher education.

"Class, I'd like to introduce our guest this morning, Stew Cohen, and he's news director of the radio stations STAR 105.5 and Y 103.9FM."

"Thank you, Miss Kilb. You are sixth-graders, right?"

They were not quite engaged and didn't answer. They were still checking me out.

"How many think I said you are fourth-graders? Show me hands. How many think I said you are fifth-graders? Hands. How many think I said you are sixth-graders, right? Hands, thank you."

While almost all knew the answer, a few had not raised their hands yet.

"How many of you didn't hear me say anything about what grade? Hands. Now that takes in everyone. Go ahead and put your hands down."

They were great.

"Okay. I have a piece of news copy from my newsroom. I'll give it to the classmate among you who can accurately tell me what I actually said *first thing* in this classroom."

Lots of looking at the ceiling, their faces given in to thought ... a smile here, a frown there ... some turning to their teacher thinking maybe she'd do something with a face or hand gesture and help them solve the puzzle.

"Raise your hand if you'd like to give it a try," Miss Kilb advised and pointed to one of her students who'd raised his hand.

"Miss Kilb, the first thing said was *thank you*."

"Would you like the copy signed?"

"Yes."

I had a magic marker.

"This was a fairly simple exercise. Good listening skills can be rewarding, as you've just seen. What is your most favorite subject in school?"

"Math." A girl raised her hand, answering my question.

"Okay. Anyone else?"

"English," a boy quickly added.

"Anyone like studying science?"

Four hands went up. You are?"

"Rita."

"You like science?"

"Yes."

"This is a classroom of connected students, I can tell. You have taken notes, asked questions, volunteered. But what happens if you are not interested in science, math, or English?"

No one raised a hand.

Then a boy in the back row raised his hand, though he seemed unsure, only raising his hand to his shoulder. Miss Kilb called on him.

"I don't like a lot of homework."

He had me laughing.

"When your teacher talks about homework, you may start thinking you might not get to go outside and play with your friends as much as you'd like after school, and you start thinking of all the things you do with your friends, and the thought of homework disappears."

I had their attention now. I could tell. Who liked to have homework take away from their after-school fun? For the most part, I've found that students are good listeners. These are students I've come across in years of visiting schools. However, school days are long, and there are subjects that some students just don't like, and they may tune in and out. I've given them an actual application applying to my life.

"Now, listen closely and tell me what I've just said," I prompted the students to super listen.

I set the scene, telling them I'd been interviewing a police officer where a crime had occurred.

"Shjrl qheish rurusnie," I said slowly and looked at their faces. To say the kids looked confused might not completely describe their response.

"Anyone?"

They couldn't guess at what they could not understand.

"I didn't know I was short in the skill of listening until I reached under police tape at a crime scene as a reporter for WYEN Radio. Standing next to a police officer and a car they had for evidence, I asked my first question of the officer."

I think I broke the barrier of what Miss Kilb's kids perceived as a lecture where they are very passive and are expected to absorb information versus where they are active and trying to be one step ahead of me in the story.

"What happened next?"

A couple of students raised their hands.

"Yes."

"The police man made an arrest," speculated a girl in the class, looking around the room at her classmates nodding their approval, though they couldn't have possibly known what the officer said.

"Yes, police investigated, and eventually found the suspect in a downtown Chicago hotel."

We moved from listening skills to writing and reading, and how all of those key subjects open the door to future success.

Young people almost always base their level of involvement in education on why it's important to them. Before they invest time and energy, they've got to have a personal connection. Miss Kilb counted on me to demonstrate how the skills I described were planted and cultivated in college and bloomed and grew throughout my career.

Finally, just some thoughts on listening skills that you may or may not know. You might think babies are good listeners. They don't have the baggage of adults, thinking of politics, losing weight, or wondering whether they have enough money to pay their bills. Babies' minds are clean slates. They seem to absorb more and at a faster rate than any other age. Picture a baby focused on mom or dad. That concentration is on a level unmatched anywhere. Wouldn't you agree this is what pure listening resembles? Although they can't communicate in sentences, babies rather quickly develop a way of gaining our attention and listening to our cues, and they respond by smiling or crying, or wondering whether they should cry and smile at the same time at something we've said. Somewhere along the way as we grow older, we lose a baby's pure focus. Outside sources like Facebook, television, radio, cell phone apps, and iPods compete for our attention. Friends need a Facebook message or a return cell phone call or texting, and we begin to feel torn in many directions. I didn't even mention competition from TV sets and radios that are left on all day and drain our attention. All these sources are pounding at our eardrums. How then are we supposed to focus our attention on anything?

Listening skills seem to deteriorate over time—an inevitable loss of focus compounded by too much stimuli. Babies morph into adolescents and users of the television remote or Xbox or PlayStation, and they play games too on their Smartphone. Where once the baby was like a sponge, gathering in all the words we'd say, the minds of adolescents and teenagers gather in the words we're saying, providing we are saying the words carrying a deep dish pizza, hotdogs, or chicken wings into their room.

Teachers in the elementary schools recognize the importance of listening skills, though I don't know how much time they spend working with their students on their listening skills. Yet the value of listening is immeasurable. Proper listening skills help students understand instruction, leading them to do the right homework assignment.

In my day, grade-school teachers were always telling us, "Raise your hand before you speak, and I'll call on you." We heard these fairly simple instructions, but we seemed to disconnect. I remember classmates and I spending way too much time in class with our heads down—our punishment for way too many times talking out of turn, ignoring simple instructions. Teachers did as much as they could by emphasizing the importance of listening skills and trying to apply the methods.

Today, I could go under the police tape, if given permission, come face to face with an officer, tune out everything around me, and ask question after question, developing the story as we moved forward in the interview. None of the time wasted. The skill of listening started for me with the *body in the backseat*.

Country Music Inn

On New Year's morning, 1978, I woke with a hangover. For a moment, watching the walls spin around my bedroom was interesting but scary, for I hadn't seen such spinning since Dorothy Hamill won the 1976 Olympics in figure skating. Though dizzy and in a haze lying on my bed, the memory of my New Year's was fairly clear, though I wished the whole experience had never happened. I needed sleep, but I couldn't stop running into the bathroom, hugging the toilet bowl.

Five hours earlier, I finished my last shot of peppermint schnapps at the Country Music Inn, ate my last chicken leg, heard the music end, and watched the country singers leave the stage, and my friends and WYEN coworkers Val Stouffer and Dan Diamond and I headed out of the bar and into the parking lot.

How do I remember one particular hangover from more than thirty years ago, and why is my stupidity worth telling you about? My night at the Country Music Inn was thoroughly disgusting, yet it shaped the basis of my attitude toward drunken driving I'd adhere to in the future.

It didn't make me feel any better kneeling at the toilet, resting my head on the seat, waiting to remove all the contents of what I ate and drank at the Country Music Inn on New Year's Eve. I only had one other experience quite this bad. I didn't make drinking to excess a habit. I was in college, living off campus, and the guys in my apartment started drinking screwdrivers—vodka and orange juice—since we had plenty of vodka and orange juice. Like a bunch of idiots, we made it into a drinking game. After that hell I put myself through, I swore off anything more than an occasional glass of wine or beer, but I never again intended to drink so much I'd feel like the room was spinning. I've held pretty steady to my conviction, except at the Country Music Inn, and the time in the early 1980s I was working for WIVS-AM in Crystal Lake, the Bellairs' radio station, that I drank more than a few cups of beer in an hour at a village demonstration run by the police department. This demonstration was supposed to show the effects of alcohol on balance and speech during a set amount of time. Panelist guinea pigs drank beer after beer. But the effects hadn't kicked in by the time the police hoped to demonstrate how wobbly the guinea pigs were that evening.

"Stew, please stand and walk on this line of tape I have on the floor."

He was a pretty friendly police officer. I had every reason to cooperate. In different circumstances, an encounter on the road might not find the officer as friendly giving a series of sobriety tests. I got up and stood where he pointed at the tape.

"Right here?"

"Yes, now please walk forward, staying as close to the line as possible."

The crowd watched. No one smiled. This was serious business. I walked on the tape, hiding seventeen inches of tape at a time by plodding down both of my feet. I walked the line. No stumbles, no slurring of my words; the audience watching this demonstration wouldn't know I drank what seemed like a keg of beer. Had we waited another five to ten minutes ... well, the audience inside the Lake-in-the-Hills Community Building might have heard someone shout, "Cleanup on aisle five!"

The police chief drove me home that night. Suddenly, in the passenger seat, I felt the full effects of too many beers. I could only clearly see the hood of the squad car, and the hood swayed back and forth, and I could see a few feet of road, and the yellow lines swayed in time with the hood; everything beyond was too dark. Now I knew the Country Music Inn experience wasn't nearly as awful as I had remembered. This time I felt much worse if that's possible.

"How are you doing Stew?" the chief asked innocently enough.

What would he do if I told him I'd been praying I wouldn't in a moment of panic roll down the car window and stick out my head? I kind of liked the squad's leather seats.

"We should be home soon," I offered this as more of a hopeful comment to myself than anything factual because I couldn't really tell whether we were much closer to my apartment in Crystal Lake. The chief pulled into my apartment parking lot, stopped the squad, said a few words I can't remember, and I got out slowly, still working on the appearance of sobriety. I woke up with a terrible hangover and swore never to do anything of the sort again, not even for the police in a demonstration exercise. That experience came six years after the Country Music Inn fiasco.

In the Chicago area of the 1970s, before Mothers Against Drunk Driving (MADD), Students Against Driving Drunk (SADD), and the Alliance Against Intoxicated Motorists (AAIM), very little attention focused on drunk driving and consequences. The number of crashes caused by drunk drivers skyrocketed in the 1970s because people didn't take the issue as seriously as they would within the next ten years. Penalties for drunk driving were that of a lesser crime. Not only did the punishment seem like a step above a slap on the wrist, but young drivers lacked a general knowledge on how alcohol impairment was a result of consumption based on body weight and time. This environment of the 1970s was the environment in which I was a new driver, but that's no excuse really for not knowing enough about alcohol and its effects on good decision making on the road.

Five minutes outside without the warmest coat, hat, and gloves, and you'd likely freeze to death the night of December 31, 1977. The snow was falling slowly, yet the temperature was dropping faster—two below as night approached, five below in the darkness, and ten below as the Country Music Inn closed at 2 a.m.

Val, Dan, and I celebrated inside the Country Music Inn on Milwaukee Avenue and Aptakisic Road in Prairie View, Illinois. We sat at a long table, and you could say the New Year's Eve party reached its peak between midnight and one o'clock. I'd been nursing a couple of beers most of the night and eating from a bucket of fried chicken, with sides of coleslaw and potato salad. Shortly before midnight, the guy next to me with the cowboy hat bought the first round of peppermint schnapps for the table, probably twenty shots, and each of us took a shot glass from the cowgirl waitress working our table. I'd never had peppermint schnapps before, but I figured we had plenty of time, and the food was still warm, and people were laughing at our table and downing shots in one gulp. Then another round came; someone else paid. Figured we could go all night eating food and having an occasional drink. Or maybe the waitress might take the food away. I swung my legs around, got up off the bench, looked at the revelers singing and drinking in the room, and then I stepped lightly toward the fried chicken at a table along the wall near the exit.

The room began spinning fast at first, as though someone turned on a ride. I wasn't strapped in and nearly lost my balance. I hadn't felt anything sitting at the table!

Stand still, I thought to myself. *Don't move; just turn back toward the table.*

Turning was tricky. I moved slowly in a tight circle, but I felt as though too much turning might worsen this nauseous feeling growing worse by the moment. The bench looked so safe that determination started replacing all the areas in my head where panic had spread. A waitress walked in front of me carrying another round of schnapps to our table. The whoosh of air behind her engine nearly floored me.

It's just steps, nothing more, one step, then another, and another to the table, I silently convinced myself. Gaining a bit of confidence on every step of the ten to fifteen on my path to the bench, I looked up at the room.

Oh no.

The walls of the County Music Inn were still spinning.

Don't throw up.

Shots of schnapps lined the table. The cowboy grabbed the glass where I had sat.

Let this ride end soon.

The bench wasn't spinning.

Shuffle if you have to, just a couple more steps.

The cowboy moved slightly to the right.

There.

Safety of the bench was mine.

Management had a closing time in mind. I didn't sit at the table for more than five minutes when I heard a waitress say, "We're closing." But the impact didn't hit me until a more powerful voice silenced busy talkers. Over a speaker where country music blared almost non-stop, someone got on and said, "The County Music Inn thanks you for bringing in the New Year with us, but now we're closing."

Someone turned off the lights in the back area, near where the band had performed, and the people disappeared at the table where I sat. Even the cowboy with the hat was gone. Now, I got up slowly, hoping I wouldn't stumble and fall or slur my words. For some reason, I tried hard hiding my condition from Dan and Val. Standing, I hoped the room could slow down before I moved further, but they headed for the door.

Can't we wait another couple of minutes and talk? I kept the question to myself.

They headed for the door. I joined them, walking to the rear door, which someone left open for customers who parked in the back lot. I walked a few steps behind them.

THE WYEN EXPERIENCE

The combination of fresh arctic air and liquor hit me hard. You know that feeling of extreme cold nearly crystallizing any moisture on your face or in your nose. I pulled a glove off my hand for a moment, reached into my pocket for my keys, put the glove back on, and kept walking with Dan and Val, until I looked down at my hand with the keys and noticed my key ring had unscrewed, and my car key was gone! This happened once before where the ring opened and a key fell out, but I found it quickly where I stood. That time, the key fell on a sidewalk and made a soft plinking sound, just loud enough to hear. This time, the key fell in the snow, and unfortunately, the key might have been five feet from the rear door, or ten feet, or fifteen feet. I had walked too far before I looked down at the key ring. Strangely, panic seemed an unlikely response. My reactions were affected by peppermint snapps.

"Guys, I think I dropped my car key. It's in the snow here somewhere."

I didn't want them thinking I was panicking, and I must have come across under control. They turned in the snow and faced me. I didn't ask them to drop to their knees and search the snow for my key, not in the middle of the night in the nearly vacant parking lot of the Country Music Inn.

"I'll find it, don't worry. Just get going."

Dan and Val were hesitant. They wouldn't leave.

"Go home, guys. I'll be okay," I insisted. They remained hesitant, but I told them I'd be okay. Embarrassment meant more in my mind than doing what made sense. I feared they'd think I was an idiot for losing my car key and for not finding it quickly, and for putting them through this futile attempt to find my key, but probably worst of all, I didn't want them knowing I had way too much to drink.

"Good-bye, I'll talk to you at work."

It's got to be below zero. I knew it was two below earlier in the evening. The wind chill probably made it feel like death. My winter coat, gloves, hat, thick socks, and boots weren't enough.

"Maybe a few more minutes out here in the tundra of hell frozen over might be okay. Got to find my key and sit in the car and ride out the spinning parking lot." Yeah, talking out loud in the empty lot might be a sign I was already going crazy.

Why didn't I just let them drive me home? I should have, yes; I can clearly see that on any day but that one, where I was stuck in sub-zero weather with my brain in a deep freeze, and the Earth was spinning. I didn't want to have to return the next day to pick up my car.

I briefly watched my two coworkers leave. I thought for a moment I'd shout their names and bring them back, but I didn't say a word. They drove away. I turned and looked at the building and then at the ground. There wasn't a car in the lot but mine. Now I realized I couldn't find help if I wanted. The place was surrounded by trees. The Country Music Inn closed, the lot was empty, the snow wasn't stopping, and even though I wore gloves, the chill had numbed my fingers, but thankfully one thing was going right; the parking lot lights were still on, but for how long?

Just talk some more, I thought. Hearing something in my ears might keep me from panicking once the schnapps started wearing off.

"I'll find my key. I'll retrace my steps, and anyway I didn't go too far, I don't think."

My fingers weren't moving in my gloves, my legs were stiffening, yet my mind was speeding through pages of what to do now, and suddenly, my page turning stopped on the S's, for Super Ball. Sounds crazy, and I thought the cold and drinks were blending my thinking into a Slurpee. But something had to give soon because the new snow was falling too fast now, covering those first steps, and all I did was add more boot prints.

Super Balls bounced so high and often pinged in unpredictably impossible angles, but they were so much fun. I used to play with the ball for hours as a kid until pieces of it chipped off all over my yard. The ball eventually fell apart completely, especially the way I played, throwing it against a very uneven brick wall in back of the house. None of those memories helped. But what popped into my head from a memorable moment did have some meaning and brought forward the last oasis of sanity. I recalled playing outside under my parents' overhead telephone wires that stretched from my home to a huge wood utility pole. I was either throwing the ball as high as I could, trying to hit the Glenview Naval Air Station airplanes flying over my house, or aiming to hit the overhead phone lines on the ball's way down from the stratosphere. I thought of one such time as a kid that was crucial to finding my key in the snow. The Super Ball I tossed forty feet in the air, as I remember, hit the lines and careened away. I heard the ping of the ball, how loud it sounded, and figured it didn't make a noise hitting the neighbor's yellow barn or his wood fence, didn't break a window on my house, the dog didn't bark next door, so I eliminated as many possible landing spots, reducing the likelihood to one or two places, and I looked and quickly found the ball in an area where I thought it landed. Could I use this Sherlock Holmes deductive reasoning on my Chevy Vega key in the snow and avoid a medical emergency in the parking lot? What choices were left? My path to the car was fairly straight from the Inn's rear door, but I realized I didn't walk far enough back to the building, and so I walked as straight as I could, looking down, covering as much of the snow as I could see, and in what felt like an eternity but was really a matter of seconds, I came across my key nearly covered in new snow. The metal head and part of the keychain hole stuck out, and I reached down. I sure didn't feel the cold air through my clothes in that moment.

THE WYEN EXPERIENCE

I put the key between my thumb and index finger, the only two digits where any feeling was left in my right hand, and I held tightly. Drama has never been my forte; however, in the forty feet between my car and where I stood freezing to death, a determination pushed blood through my heart and into my veins, reenergizing my steps, propelling my feet forward to my car.

Was the car going to start? That didn't even cross my mind until the door key barely moved in the lock.

I squeezed into the car and felt the coldness of the seat on my frosty pants, so I didn't know which was colder, the seat or my pants, but my legs were tingling anyway. None of it mattered now. All that mattered was the sound of a Chevy engine.

I pumped the gas several times—then waited—and turned the key in the ignition. The car started! I pumped the gas again, and again, and again. The motor's case of bronchitis cleared, and I turned on the heat.

Staying in my car overnight never really was a serious consideration, though an hour in the car might by okay. But I had only half a tank of gas, so I couldn't let the engine run too long. The Vega was not exactly a car for winter driving or for starting in brutally cold weather.

I checked my watch and figured on a half hour sitting in my car, staying warm, and hoping I would not fall asleep. I turned on WYEN and listened to the radio for a while. I thought I felt well enough and opened the window slightly, thinking the blast of air might keep me awake, and I repositioned the front seat, moving it closer to the steering wheel and windshield, believing I'd see better out the window. I just wanted to get out of there, go home, and sleep in my bed. Driving on Milwaukee Avenue was fairly easy, just a straight shot past Palwaukee Airport, until the road reached a fork, splitting into River Road and Milwaukee Avenue, and I'd simply stay on Milwaukee Avenue to Golf Mill Shopping Center. Unfortunately, I hugged the curve to the right, not left for Milwaukee Avenue, and followed the road around and realized I was now on River Road. For a moment, I couldn't figure out what happened or where I was going. Things didn't look familiar, but I just kept on and came to Dempster and River, and suddenly I felt a great relief because I knew my alma mater, Maine East High School, was nearby, and I turned, heading for Harlem and Shermer and home.

I've felt a self-imposed penance for bad decisions in the 1970s even though drinking and driving had not yet been uppermost in the American psyche. I still feel I owe a lot to God for getting me through that without hurting myself or someone else.

Every holiday, I interview the traffic safety coordinator for the Illinois Department of Transportation. His message is consistently on point, "You Drink and Drive, You Lose." The campaign has helped reduce the number of crashes caused by drunk drivers. I am also providing serious air time on the seatbelt campaign, Click It or Ticket and Operation Click, getting teens to wear their seatbelts and drive responsibly. I've interviewed AAIM, Alliance Against Intoxicated Motorists; MADD, Mothers Against Drunk Driving; and SADD, Students Against Drunk Driving now called Students Against Destructive Decisions. I talked to Robert Anastas, the SADD founder, about how he created SADD after the death of two students, and I applied at one point for a position with MADD as its national public relations spokesman. I won the Associated Press Illinois AP Broadcasters Association 1984 Contest Continuing Series on Drunk Driving.

How I've changed is very much how many people have changed in their thinking on drunken driving because of the top-of-mind campaigns. In the 1970s, I never thought of using a designated driver because we did not talk about a designated driver, nor did I think of the consequences of drinking and driving because it wasn't a high-priority public service. The devastation caused by drunk driving was just a horrible film you'd see in a high school driver's education course. Now we read and hear messages on drunk driving and its consequences as an integral part of our daily lives.

I'll do all I can to spread the message of safe and sober driving … that's my pledge.

Chapter Thirteen

NEWS IS THE COOLEST

Stew Cohen

"Stew, play that record there." Rob pointed at the turntable on the right of the console. "It's cued and ready."

"Okay, Rob."

A stack of records on a table had grown by a Carpenters album, but Rob scooped up the whole stack on his way out of the on-air studio.

"It's all yours!" he reminded me.

"Thanks, Rob."

This wasn't the time for conversation. Reynolds had a five-minute break. He used the time returning albums in the WYEN record library and pulling out music for the next half hour. UPI News for the next sixty seconds gave me enough time to make sure the copy pages were in the right order and I didn't miss any potential landmines of spelling mistakes and grammar.

Rob only had to find a song for one of the turntables ... so I could hit the button and start the music at the end of my news, but he had a habit of standing behind me at the very moment I touched the start button on the turntable because I had a history of accidentally banging into the tone arm after I spun the chair around to leave the studio.

I was in the room alone, yet I knew Mr. and Mrs. Walters and my parents and grandparents were listening. Mr. Walters expected a "broadcast tour de force." My parents settled for something less than a "tour de force," while my grandparents didn't so much care whether they heard perfection; they just were happy they woke up that morning. Actually I did something for my grandmother. In working at WYEN with its fairly far-reaching signal, my grandmother heard my newscasts from where she lived in the old Surf Hotel in Chicago. Grandma K felt so much pride we had actual conversations about my radio experiences.

I described my anchoring as though I were a beginning watercolorist. To use the medium correctly, I had to mix the color just right with water. I lacked artistic techniques though. In broadcast terms, I lacked phrase articulation, voice character, and power of delivery. Seems in a business where you're fighting hard for every listener, having a less than professional approach can only hurt. Yet I had an opportunity to enhance the station's musical strength by delivering a vocal presentation demanding listeners' attention, and I worked hard to make this happen.

Several of the very best radio news people at WYEN had comparable artistic qualities of an Andrew Wyeth of the 1970s—true artists, not with a brush, but with a microphone, typewriter, and recorder. During the WYEN years, John Watkins, Dave Alpert, and Bob Roberts worked on stories and anchored newscasts. Each of these broadcasters was extraordinary at WYEN and throughout their careers because they combined hard work with tremendous talent for writing, interviewing, reporting, and anchoring. In every sense, they were true artists of broadcast news. They placed great value on the words they read. Every word had value from "A to Z," and every word was necessary. As they developed word pictures and drama in each story, one might sense a newscast was their art gallery. Each story had a beauty unto itself, but overall, the newscast informed and entertained sufficiently to keep the listeners there.

Beginning broadcasters tend to put all their concentration on "not screwing up." They try so hard to pronounce the words correctly, read without making a mistake, and write clearly. Watkins, Alpert, and Roberts emphasized newscast preparation but not to the point of overemphasis. Why put too much pressure on "not screwing up," because in their experience, that's exactly what happened. More importantly, they understood communication of words. They had the skills for changing vocal variety, dramatic pausing, and articulation. Does every sentence sound the same? Is the story about a centenarian's birthday read the same as a story about a murder? Each receives special treatment. They properly conveyed different feelings because they understood each story was different. As a news writer and anchor, each communicated a whole palate of feeling that connected with the listeners.

I was blessed to have known and worked with Watkins and Roberts,

THE WYEN EXPERIENCE

and I was an appreciative listener of Alpert's news on WDHF and WMET, Chicago. Each broadcaster could cover any type of event, and they knew how to weave information they gathered into informative and entertaining stories. It didn't much matter whether their deadline was five minutes or five hours for a story. All three saw things others could not see as well, and they'd create easy to understand word pictures. The one constant in their abilities for gathering news was their way of honing in on the *why* of a story. Everything that happened had a reason, and each of them knew how best to *uncover* the reason from an interview.

While chapter 9 looked at Roberts's story, this chapter focuses on Alpert and Watkins and the rest of the WYEN news department. But first I'd like to elaborate on the secret of their success. I mentioned they honed in on the *why* of a story, but for this to compel listeners to choose them among the many anchoring news depended on how they harnessed human emotion. They didn't wait; they pursued stories. Alpert, Watkins, and Roberts were aggressive in their approach, however, not exactly aggressive as a bull might be in the *Running of the Bulls* through the streets of Pamplona. While always rushing, these bulls didn't gore anyone. Instead, they instinctively understood the emotional aspects of the stories they covered. Reliving how a firefighter saves a child in a burning home, or exposing the struggle within people to approve a school referendum knowing they couldn't afford the additional taxes are but two examples of what radio can do well in its appeal to the ear and the heart. Alpert, Watkins, and Roberts have had a history of going beyond the headlines; unfortunately, few stations today are letting news people venture beyond the scope of headlines. At WYEN, Mr. Walters gave us great latitude in developing features and hard news. One such example occurred early in my development as a news reporter at WYEN.

On September 13, 1976, a Chicago police officer lost his life on duty. Willie Lewis pulled out a .44 magnum and fired once at close range, killing twenty-six-year-old Patrick Crowley.

Crowley had made a name for himself in his police work conducting undercover drug operations. This one though had an outcome that led to nearly a thousand years in prison for Lewis. Crowley and two other officers were about to remove a narcotics dealer from the streets, but as Crowley and the other officers discovered Lewis, he fired at Crowley's head. Lewis didn't miss from four feet away. His rap sheet now had on it: cop killer, narcotics, gun violations, robbery, and battery. Crowley left behind a wife and two small children.

Back in the 1970s, I produced a law enforcement program on WYEN. I'd record interviews with suburban police chiefs, police liaisons, Officer Friendly, and others on various subjects involving safety and other issues.

I contacted a Park Ridge police sergeant for an interview on some program in the community, and he offered me an opportunity to ride with him in a procession of squad cars for Crowley's funeral. I signed a waiver of injury card. I didn't bring a recorder. I didn't think it was appropriate. We did not join fellow officers in the church where the services were held. Instead we sat in the Park Ridge squad in a line of police cars and waited for the procession to start with police cars from Mount Prospect to Oak Lawn. This was an extremely long procession on major Chicago streets to the cemetery for a full ceremony. I jotted down notes, formed a poem I believed gave it dignity that a quick news story might not convey. By the end of the procession, I finished my poem.

> *He was doing his job, a narcotics bust.*
>
> *He kicked down a fence, shouted a warning, and rushed up a short flight of stairs.*
>
> *Four shots rang out.*
>
> *One stopped him.*
>
> *Chicago policeman Pat Crowley.*
>
> *A procession of police cars covering four miles of road; its destination, a church where more than one thousand fellow policemen paid their last respects to one of their own, a brother.*
>
> *A line of police cars stretching as far as the eye could see gave evidence of a great togetherness; a family with a sameness of purpose.*
>
> *This is the Brotherhood of Policemen.*
>
> —Stew Cohen, WYEN Metro News 1976

The use of the word "policeman" was common before gender specific references disappeared from our vocabulary. Today, I'd say police officer, rather than policeman or policewoman.

THE WYEN EXPERIENCE

On Bill Wilkinson's last day, I thanked him for his help in the newsroom and his tips for anchoring news in the WYEN on-air studio.

About halfway through my time at WYEN, and with the departure of Wilkinson, my news director, for employment in management at Brown's Chicken, Mr. Walters chose to name me news director. What I had liked about Bill was his down-to-earth approach. He did not come across as slick or somehow stylized; his was the "meat and potatoes" voice your neighbor might possess. We worked well together. Bill was more a coworker than a boss. The job of bossing was left to John Watkins, a veteran of radio news. Mr. Walters learned that John had interest in returning to WYEN after a stint at a Washington DC radio station. Watkins worked for close to six months at WYEN, helping Bill and me as his afternoon and evening anchor reporters. I guess he believed we needed a lot of his guidance because he sent inter-office correspondence to us almost more often than I watched *Rocky* and *Saturday Night Fever* in the theater and on television over the years.

Watkins's knowledge of radio news was a blessing for our news department. Although I appreciated Wilkinson and his news direction, he did not have the experience Watkins brought to us. John produced a lot of office correspondence, but Bill and I really needed his direction, guidelines, and instructions. John listened closely to us on-air, better than most of the consultants I've seen over the years. He spent time actually participating in the effort to deliver news from our newsroom and knew what one could expect and knew specifically what hands-on direction the news anchor needed for improvement.

The operation of news departments has changed over the years as technology has advanced. But in the use of paper documentation, Watkins was a master at WYEN, and in December of 1977, he produced specific instructions.

Cart labels must carry a certain amount of information. The label should have the following information on it: the date, name of the person on tape, a notation of the story slug, outcue, and length of the cart.

This information is necessary since we'll all be leaving tape for each other. When filling out labels, you should use either pen or felt tip marker, and the information should be printed. This will make it easy to read.

The labels for news carts should look like the following:
Bufica/Weather
12/7a ... your carburetor :21

The label will help an anchor spot carts faster, will let him know which pieces are or will soon be outdated, will allow the anchor to time his newscast by knowing how much time tape will take. (In the very near future, you can expect it to become standard operating procedure to time out newscasts exactly ... if it calls for a three-and-a-half-minute cast, it will be three and a half minutes, not 3:15 or 3:45. Also these labels will be needed in case I have to save actuality or voicers for award presentations or for possible "Year in Review" shows.

Our labels, our voicers and/or wraps should contain a standard outcue. Our standard outcue/tag will be Bill Wilkinson, WYEN Metro News. That tag will hold for in-house material. If you cover a story on the street, the tag should be "At City Hall, Stew Cohen, WYEN Metro News." Stringers should also follow these tags. That would mean that the Medill stringer should sign off "In Washington, Joe Blow, WYEN Metro News." (If the stringer is on Capitol Hill, then Capitol Hill should be the locator in the tag.) A standard outcue should be noted as SOC on our labels.

When you are through with each newscast, you should put your copy in order and staple it together. Before you staple it together, you should type a cover sheet for the cast. This cover sheet is simply a sheet that is to be stapled on top of the newscast copy. The cover sheets should have the following information on it:
Date
Time of Cast
Anchor's Name

A week later, Watkins sent another inter-office correspondence to the news department reminding his team of standard operating procedure. The procedures include the "chances of dry weather." The dry weather chances were a Mr. Walters signature technique, something he absolutely demanded of his news department. He believed in ending the newscast with a very positively worded weather report ... so dry weather chances, even if only 20 percent, represented a positive view on the weather in the owner's estimation. Watkins provided us an example of a weather report. Projected high and low temperatures were used rather than a specific temperature in the forecast. That way Watkins knew we would not fall into a trap saying "a high of forty-two" and then read the current temperatures and say "forty-four at O'Hare Airport." I've actually heard this on a Chicago radio station, so the anchors are caught not readjusting the high temperature. Nothing wrong with saying "high in the low to midforties."

> *WYEN Metro Forecast*
>
> *Today, cloudy. Dry weather chances 70 percent. High in the low to midforties.*
>
> *Tonight, clearing and cooler, possible frost. Low in the lower thirties.*
>
> *O'Hare 41*
>
> *Midway 39*
>
> *The Loop 40*
>
> *I'm Bill Wilkinson for WYEN Metro News*

Every news director or program director shapes the way the staff will present the news. This is their signature on the station, and they spend some time working out the best way to deliver the newscast in the most interesting manner, yet remaining concise and informative. Watkins was no different from others before him in giving his spin on the format of the cast. From being in the business for close to forty years, I can tell you no one has developed the perfect news presentation. Each format has positive and negative aspects, though the negative can be masked by great news delivery and writing style. Imagine the news anchor preparing a newscast with the Watkins formula and then pushing through the cast to the end. He'll make the cast as smooth as possible, emphasizing any mention of WYEN. His voice will sound pleasant on the introduction, news sponsor, and conclusion of the news, and switch a more serious delivery through the hard news. He'll provide differing lengths of stories for pacing and then round out the cast with a feature story, brief sports, and weather, ending with a very authoritative reading of his name and the station call letters. On December 22, 1977, Watkins wrote about another key component. He was well aware of how listeners depend on radio for not only great music and information, but on the time. Time checks help listeners stay on time to go to work, school, or wherever they need to be any time of day.

The FM 107 News time line in our newscast should come before the last news story in the cast. This will separate our time checks. All news departments need a format for their newscasts and, the best are the casts that flow smoothly from one element of the cast to the next, such as from the lead-in to the sports and weather.
    :59:55 WYEN ID
    01:00 UPI World News
    02:00 UPI runs a spot
    04:00 WYEN Metro News
    04:30 Local News Sponsor (Woodfield Ford/Lattof Chevrolet)
    05:00 WYEN Sports and Weather

For all the office memos from Watkins, one thing he didn't write. He didn't describe how we settle into the seat and get into the mind-set of delivering a newscast. The correct procedure was a given. Watkins expected his news department to know how to perform professionally the moment each of us walked into the main studio.

The news anchor entered the main studio usually five minutes before the newscast, enough time to let the announcer get up and leave the room, and then the anchor would sit down and review his or her newscast, make sure the carts with sound bites were in order and in the cart machine and the volume on the board was turned up to play at the right audio level, and the "spot" billboard (Woodfield Ford or Lattof Chevrolet) was ready to read from the copy book in the studio. Then it was time for plugging in the headphones, adjusting the volume, and listening to the outcue on UPI, "You're listening to UPI World News." Make sure to have one hand on the "key" to the volume control or just turn the volume down quickly and make sure your microphone level is up and the "key" above the volume pot is flipped to the right for program, not to the left for audition. Remember, breathe in and relax. You are ready for WYEN Metro News.

    Good morning/afternoon,

    It's _____degrees under _____skies in Chicagoland at _____:02.

    I'm _____reporting for WYEN Metro News.

THE WYEN EXPERIENCE

Heart of the newscast; local, national/international, state, feature (soft story)

(Time check to spot of local advertiser) 107 FM News time _____

(Spot)

WYEN Metro Sports

(Brief report)

WYEN Metro Forecast

Today,

Tonight,

O'Hare Airport

Midway

The Loop

At _____:05, I'm _____ for WYEN Metro News.

Watkins changed our schedules. Flexibility is key in this business. Watkins would let me know in another memo.

Stew will be moving out onto the street to give us some coverage. It would be wise for Stew to see that he is back at WYEN no later than 5 p.m. since he will have to pick up the 5:35 traffic. Cohen—Monday through Friday, 1:15 p.m. to 7:15 p.m. Stew's first air assignment will be 5:35 p.m. traffic. On Sunday, Stew will hold down 8:30 a.m. to 1:30 p.m. His first cast Sunday will be the 9 a.m. Stew will also leave rewritten material with sound for the weekend jocks.

In late 1977, Watkins told his news department to start keeping a log of everything done in the newsroom. It sounded like a lot of extra work to write down everything, but this being a communications business, the news team should go all out and keep everyone in the loop, Watkins thought. A state representative or a school board member might call, and the anchor should be prepared because he understood what he'd ask them as written on the log. Was an interview necessary, or would we ask when the best time was for a return call? Was the person asked to clarify something he was asked earlier in an interview? So many questions could be answered by just reading the radio news log or journal. Today, e-mails serve a similar purpose. Watkins gave a fancy title to another component in the newsroom.

The Radio News Desk Log should reflect minute-by-minute developments in the newsroom. On this log, you should indicate the time you interviewed on a story. What callbacks you are expecting. What stories broke and what is being done to follow them.

This log will come in handy for documentation on reaction speed. What the news team did and are doing on a story, and it will also let the following anchor know what has happened and what is happening in the newsroom.

This log should also be used to alert the Manager of News and Information to technical problems encountered by the news staff, what, if any, problems there are with commercial copy that the news department has to deal with … (If tags run longer than :10, I want to know about it!)

The log should also reflect what tape we are getting, when WMET feeds a story and sound bites, the story, who fed, and brief summary of the cuts should be logged. If we get tape from any other outside source, that is to be logged.

We should also begin keeping an audio log. This shall simply be a sheet of paper listing the name of the person on the tape, a brief description of the tape cut, the outcue, and how long it runs. Also, cuts should be numbered sequentially, and numbers should be assigned carts, and the number of the cart should be noted on the audio sheet. All tape that pertains to a particular story should be kept on one sheet of paper. If you have more than one person on a story, each separate interview is kept to one sheet of paper and attached to the rest of the copy. The best way to deal with City News Bureau, CNB, and other wire copy is to keep the latest copy on top. All copy on a story should be saved with the latest wire copy stapled to the top of the bunch. This will allow you to have the last and freshest copy on any story available at a glance, with background right there also.

Watkins was soldier savvy. He operated the news department as though he were a general and we were his captains. He spelled everything out in his memos, and expectations were high.

With training from Watkins and Roberts, I felt fairly confident to direct my news staff of Wally Gullick and Beth Kaye. Wally's dream job was his close connection to National Football League Hall of Fame Chicago Bear Walter Payton. Very early in Payton's NFL career, I interviewed Payton. He was very shy, just finishing his first year in the NFL, and the Chicago Bears were learning they'd struck gold in this running back. Before I taped my interview with Payton, Allen asked me to convince Payton to say a few nice things about Wayne's show. Allen hoped to pop Payton sound bites into the show. I was now apprehensive asking Payton to do Wayne a favor, and I wasn't sure I could hold my own with Payton. Since I couldn't concentrate on the interview, I asked outright to accommodate Allen as we sat across from each other in a small room inside Woodfield Mall's main office.

"No." Payton shook his head briefly, a signal for me to move on with the interview.

"What can we expect from the Chicago Bears this season?" I said, slightly relieved to have the Allen agenda over.

Very quietly, Payton responded as though we were in a library and the librarian had strictly forbidden talking. Payton had his head down. Maybe he was thinking. I believe he was only starting to get used to the media. He picked his head up and stared at me.

"We're looking good on the offensive side. Defense is ready too." Payton's voice indeed sounded sweet, and I realized this was a youthful innocence he projected. While his voice was sweet, his open-field elusiveness earned him the nickname "Sweetness."

I also realized my sports interviewing skills lacked a solid foundation in Bears current facts that I could draw from for this interview. Payton did not open up, and I never hit on something to bring a long answer out of him.

Payton's autograph session brought hundreds of his fans to Woodfield Mall. Kenn Harris waits for an opportunity to interview Sweetness.

My five minutes with Payton were five minutes with one of the best football players I'd ever seen and the best player in the era in which he starred for the Bears. Though I had nothing insightful, I never forgot how Payton impressed me as a very quiet person, kind of shy, almost painfully so, in a business that lives in full-blown publicity and everything is on a very large scale. I thought, *Unless he changes, pro football might eat him up.* I was wrong.

His performance got better and better and spoke volumes for Sweetness. Payton died November 1, 1999. He was only forty-five years old.

Wally Gullick

This man was born to do sports. He could talk about eating a hamburger, and I'd think he was describing the seventh game of the World Series. Gullick worked for WYEN Request Radio and NFL Hall of Fame Chicago Bears running back Walter Payton.

WYEN was Wally's first full-time job out of Southern Illinois University in Carbondale. His friendship with Mike Roberts (formerly of WYEN, the long-time overnight announcer at WTMX-The Mix FM, Chicago) helped get "his foot in the door." Gullick interviewed with Mr. Walters and started in 1978 in the news department. Though he worked hard in news, his passion was sports and remains sports to this day. However, quickly after starting at WYEN, Wally recognized WYEN sports was not a priority. Wally handled himself very well in news with his energized newscasts. He managed to bring more interest to sports through his very enthusiastic coverage of sporting events. One of Wally's efforts was his coverage of Dave Corzine, the NBA basketball player from the northwest suburbs.

ARTIS GILMORE, AN NBA ALL-STAR IN FOUR OF HIS SIX YEARS WITH THE CHICAGO BULLS AND A PLAYER WHO WOULD EVENTUALLY BE ELECTED TO THE NBA HALL OF FAME, TALKS WITH WYEN SPORTS DIRECTOR WALLY GULLICK.

THE WYEN EXPERIENCE

Wally put WYEN on a number of media lists for pro teams in Chicago. He went to the Chicago Bears practice facility in Lake Forest, met George Halas and Pat McCaskey, and was later steered to former player Doug Buffone, and eventually to Payton, a football legend. Wally became involved with the Entertainment One Group, the management company that ran Payton's food and beverage establishments. Besides securing sponsorships support to underwrite special events, Wally had the privilege to serve as emcee for the events, and for the first time since he left WYEN, he had a microphone in his hand again, and this time in front of live audiences.

"Once you are behind the microphone, it's in your blood, and working a live crowd with high profile athletes gave me that kid in a candy store feeling."

From 1989 to 1996, Wally marketed public relations for Payton and his nightclub/restaurants in the suburbs and downtown. He'd handle special events and corporate sponsors and underwrite the appearance of these athletes. Wally says Payton provided instant credibility … so he could get on the phone with an agent of a Dan Marino (former Miami Dolphins quarterback great), for example, and get an immediate response.

During Wally's professional news coverage at WYEN, he received a letter from the head of NORTRAN, the north suburban public bus company. Dated December 21, 1978, the letter was addressed to Jerry Westerfield, VP of Walt-West Enterprises.

> *Dear Jerry:*
>
> *Just a short note to let you know that the WYEN news team, in my opinion, is the most professional of any of the broadcast media that covered the recent strike by drivers and mechanics. Stew Cohen and Wally Gullick gave accurate, balanced reports of the NORTRAN situation. Wally Gullick did an editorial piece which was objective, thoughtful and accurate. He took the time to research the facts, double check them, and had the courtesy to call and indicate when the news analysis would run. This is not to say I agreed with all of his analysis, but it certainly was a professional job.*
>
> *Sincerely,*
>
> *Joseph DiJohn, Executive Director and General Manager*

In his stint at WYEN, Wally was involved with two huge stories in the 1970s—the American Airlines DC 10 Flight 191 plane crash in May 1979 and the Gacy saga. Wally remembers editing tons of tape from Bob Roberts for sound bites for his newscasts and recording Bob's voicers on carts for replaying.

"I had been on vacation at the time of the crash, but I flew home the day after and remember walking into the newsroom amid all of the chaos and just jumped right in."

As for Gacy, Wally anchored news casts and special reports and assembled and edited the tape. He did voicers for other radio stations that called from out of town and wanted to localize the story.

Gullick submitted in my interview with him, "The radio business is such a unique fraternity because you spend so much time with these people. The friendships and relationships are special. WYEN was a family from the receptionist to the announcers to the sales people. A ton of quality people and talent worked at WYEN. I was glad to be associated with it for those three years."

Gullick's years working for Payton were the highlights of Wally's career.

Dave Alpert

I used to listen to Chicago's WDHF-FM and WMET every morning and hear Alpert anchor the morning news. The word *consummate* best describes the broadcast news approach Dave practiced in his professional career. He fleshed out stories with serious investigative effort, wrote with a creative flair, and read news with an unmistakable voice. Alpert showcased his writing and voice before a nationwide audience. Just as it made sense to let Roberts tell his own story in an earlier chapter, I felt Alpert should do the same here.

In 1974, the United States was in the throes of the Arab oil embargo. OPEC, the Organization of the Petroleum Exporting Countries, had cut off oil shipments "in response to the US decision to resupply the Israeli military" during the Yom Kippur War. Gas prices skyrocketed, and President Nixon requested service stations to voluntarily not sell gasoline on Saturday nights or Sundays. Then the government imposed the infamous odd-even system. If your license plate ended in an odd number (or was a vanity plate), you were allowed to purchase gasoline only on odd-numbered days of the month, while drivers of vehicles with even-numbered license plates were allowed to gas up only on even-numbered days. I recall the long lines at service stations as I drove to and from work at the WYEN studios in Des Plaines, where I was news director, from my parents' home in Morton Grove. As the embargo dragged on, gas prices continued to rise, and supplies became more critical. There were rumors the government was preparing to institute a more onerous rationing plan.

THE WYEN EXPERIENCE

One morning, I led my newscasts with an item that would garner more publicity than anything before or since. National publicity! I broke the news that gas rationing coupons had already been printed. I had seen them with my own eyes. The UPI wire service picked up the story, and the newsroom phone was soon ringing off the hook with calls from other reporters wanting more information. The story appeared in newspapers across the country, all attributing it to "the news director of a suburban Chicago radio station, WYEN." Bill Curtis interviewed me for Channel 2's evening newscasts.

The government wouldn't confirm the existence of gas rationing coupons, but they never denied it, either. No such coupons were ever actually issued, and even I had almost forgotten all about it by 1984, when the New York Times *ran this item:*

> WASHINGTON, June 1—The Energy Department is shredding 4.8 billion gasoline rationing coupons that were printed but never used in the Arab oil embargo a decade ago, a government spokesman said today. "It cost millions of dollars to have the coupons printed and stored at the Pueblo Army Depot in Pueblo, Colo., where they are being destroyed and buried, officials estimated. "We are in the process right now of shredding these gasoline rationing coupons," said James Marna, a spokesman for the Energy Department. "They are unusable. They were never even printed with serial numbers." The Federal Energy Administration, now defunct, ordered the coupons printed in 1974 and 1975 when oil supplies from the Middle East were cut off. The coupons are being shredded and buried because they could trigger dollar-bill change-making machines.

My first connection with WYEN was in 1969 though Mr. Walters filed an application for a license from the Federal Communications Commission in the early 1960s. But it was 1969 when I first discovered another radio station: WEXI in Arlington Heights. I blasted the station over the public address system at Niles North High School in Skokie every day before "first bell" and my reading of the morning announcements. And I applied for a job there. WEXI Program Director Ray Smithers signed the rejection letter.

Fast forward to 1974. I had just returned from Florida after attending college, working as a "copy boy" at the Miami Herald, *doing news part-time in the dying days of legendary Top-40 station WQAM and as news director at a 250-watt daytimer in Homestead. I found WYEN while flipping through the* Yellow Pages *in a phone book, under Radio Stations and Broadcasting Companies.*

"What the heck is WYEN?" I wondered. "The station wasn't here when I left for Florida!"

I called the station, told them I was looking for work, and learned they needed a news director. Some guy named John Watkins was leaving for a gig downtown. Our paths would cross again. In fact, John and I would become good friends. Anyway, I got the job, and my first on-air appearance on WYEN was doing a traffic report during Smithers's afternoon show! Ray left shortly after I arrived, and I never had the chance to tell him how he turned me down a few years earlier. Oh, did you know WYEN's transmitter and tower were in Arlington Heights (at the southeast corner of Dundee Road and Arlington Heights Road), practically adjacent to WEXI? The pay wasn't great, but it was a full-time job that allowed the radio staff to work in the Chicago market though the signal wasn't exactly saturating Chicago. WYEN operated as a suburban-oriented Adult Contemporary station, with not much signal into the city, even with its 50,000 watts. Working there gave me an opportunity to hone my broadcasting and journalism skills, and opened the door to unexpected adventures, unimagined opportunities, and long-lasting friendships.

It was there, at Request Radio, that I met a very young Rob Reynolds. After living with my parents since moving back from Florida, Rob and I shared an apartment. A few years later, I would be best man at his wedding. I never had the privilege of working with Stew Cohen, the driving force behind this book; we met for the first time nearly forty years after I left the station, at one of Rob's infamous Christmas parties. My former coworkers Garry Meier and Chuck Hillier were there, too.

Watkins had already established a strong local news operation at WYEN. When John moved downtown to WDHF, which had just been purchased by Metromedia, I inherited his Rolodex and fire turnout gear. John was a fire buff, and I was, too, though maybe a bit less gung-ho. But big fires made big stories. There were some memorable ones in the northwest suburbs, including the blaze that destroyed the original Arlington Park Racetrack, another at the sprawling Maryville Academy orphanage in Des Plaines. John and I remained friends as his career took him to WMAQ, KGO in San Francisco, WRC and the new RKO Radio Network in Washington.

In November 1974, we fielded stringers throughout the suburbs and provided wall-to-wall coverage of the midterm elections.

"The most comprehensive coverage on suburban radio," I think the promo said.

Reynolds manned the controls, switching in and out of the UPI Audio national feed, while I anchored local returns alongside our political analyst for the night, Forty-Fourth Ward Chicago Alderman Dick Simpson. I'm still not sure how Mr. Walters put that deal together, but I'll bet Simpson got paid more for that night than I made in a month. Then again, maybe he just did it for the exposure. After all, he'd be running for reelection himself before long.

While working at WYEN, I was also a stringer for UPI. Most of the stories I filed, whether for the wire or the audio service, were pretty routine. But I do recall being asked to cover and provide a live audio feed of a speech in Chicago by President Nixon, at the height of the Watergate scandal, at which it was thought he might give an indication of his intention to resign. He didn't mention it that night, but he did resign a few weeks later.

When Jimmy Carter entered the Democratic Party presidential primaries in 1976, he was considered to have little chance against nationally better-known politicians; his name recognition was just 2 percent. Very early in his campaign, Carter spoke at a synagogue in Skokie, which I covered. Not surprisingly, it was easy to get an interview.

Our proximity to O'Hare offered a potential wealth of interview opportunities, but I never really took advantage. Even closer than the airport was the Midwest Region Headquarters of the Federal Aviation Administration in the O'Hare Lakes Office Plaza where our studios were located. Making friends with some FAA workers would pay off a few years later. The crash of American Airlines Flight 191 on May 25, 1979 was the most deadly commercial aviation disaster in US history, and while I doubt I was actually the first to broadcast an interview of an eyewitness, as some people have insisted, I did get it on the air pretty quickly!

I've never been much of a sports fan, but WYEN evening announcer Jon Zur and I twice covered the Indy 500 together. I covered the World Hockey League championships at Randhurst, and Northwestern University football games at the old Dyche Stadium in Evanston. Our northwest suburban location gave me unprecedented access to Arlington Park Racetrack. Like a lot of others at WYEN, I had my turn at recording Tony Salvaro's "Harness Racing Wrap-up" when he called in every evening.

Alpert touched on his career track, paying attention to the broadcasters he met along the way and the major stories he covered. Alpert left WYEN in October 1975 to join Metromedia's WDHF, which became WMET, where he was news director until 1981. He was one of the original news anchors at Satellite Music Network in Mokena, Illinois, and moved to New York in December where he spent the next seventeen plus years as a news writer, editor, producer, and sometimes reporter at ABC News Radio, covering everything from Live Aid to the first Gulf War, political conventions, space shuttle launches, and the Atlanta Olympics bombing, and toured with the Rolling Stones. Dave also interviewed rock stars for the revival of "In Concert" on ABC TV, and did news for the "Jay Thomas Morning Show" on WXRK K-Rock. He moved to the West Coast in 1998 as executive producer of entertainment news coverage at Westwood One Radio, and in 2000 rejoined ABC News as a Los Angeles based network correspondent until September 2010. Alpert is now anchoring custom local newscasts for radio stations across the country from his home studio in California.

ALPERT FINISHES TYPING SPORTS AND WEATHER IN TIME FOR ANOTHER WYEN NEWSCAST. THE WYEN ROTARY RED HOTLINE PHONE IS WITHIN REACH IN CASE SOMETHING MAJOR BREAKS.

Beth Kaye

WYEN in late 1978 had a need for another news announcer because of Wilkinson's departure. Although I didn't interview anyone to fill the opening, Mr.Walters hired a young woman with a few years of news experience. I met Beth, and she seemed very professional and ready to work hard. She was given a shift for news and had to also provide voicer reports for me in the morning. I quickly realized I had trouble pronouncing her last name, "Krusich," so I asked Allen to help me decide on a new last name for her, and Wayne and I came up with Kaye, which was very acceptable to me because I called my grandmother "K" as in Grandma K, short for her last name, Klimboff. The K as in Beth Kaye apparently stuck because several years later I heard Beth on Murphy in the Morning on Q101-FM. She was a regular traffic contributor on the show and went by the name Beth Kaye.

Allen remembered Beth as very professional.

"I loved having her on my show doing the news. I couldn't thank her enough for helping to raise funds for my favorite charity, the Cystic Fibrosis Foundation."

Beth had been another outstanding female broadcaster at WYEN.

Carmen Anthony

Mr. Walters had a way of hiring people and then letting his managers know. In hindsight, I appreciate the fact that he took the financial risk and expanded the news staff rather than reducing its size. Radio, television, and newspaper news people have seen their numbers caught in budget trimming over the years. Mr. Walters hired Anthony for the morning shift. Carmen had a long history in broadcasting on WJJD-FM and with the Stock Market Observer on WCIU Channel 26. I was excited the news staff could work side-by-side with a veteran. Anthony reminded me of my college radio professors, some from another era of radio. They had what we'd call *golden throats*.

Nick Kumas

Kumas contributed directly to the book. He asked to reflect on WYEN and on radio in general.

I worked as morning news anchor at WYEN in 1976, an exciting election year. I had an opportunity to cover the presidential election as well. It's sad that the number of individually owned radio stations like WYEN is down to practically zero! The Communications Act of 1996 destroyed the opportunity both for potential owners as well as up-and-coming on-air people. Someone like Walters could not now own a radio station. Not only is radio programming generic and cookie cutter, but the real loss is that America is hearing basically one editorial voice, neither Democrat nor Republican, but a power-that-be voice that manipulates and destroys!"

While I may share some of Kumas' views, he and I certainly shared the knowledge of radio as a fun, challenging, always changing business, where we as talent got to do so much and meet so many people. Where could I one day interview Shirley Temple, the next day cover the John Wayne Gacy story, and the next day play Ping Pong with sports legend Bobby Riggs?

Chapter Fourteen

COULD THEY TAP DANCE?

America's Sweetheart

Gene Randall, CNN's national correspondent and fill-in anchor, saved me from serious embarrassment one day, but he probably wouldn't know this because I didn't make a big deal over my almost "foot in mouth" moment with Shirley Temple Black at the Ambassador West Hotel. My meeting with Randall in a Chicago hotel elevator was purely accidental but unforgettable, mainly for the reason we came together—to interview America's Sweetheart.

This was a summer day in 1976, and Wilkinson had given me one of my first assignments at the suburban radio station.

"I have a really good interview for you. Shirley Temple Black!" Bill's friendly tone underscored the softness of the story.

"Shirley Temple?" I repeated, trying to figure out how far Bill would go to suck me into really believing I'd actually step into a room with Shirley Temple.

"She's downtown through this weekend, and I have the contact name for Mrs. Black. Just call the contact and make sure Alpert and Roberts get a sound bite or two from your interview."

"Sure," I said, kind of unsure.

My mother and father were fans of the child star and had enjoyed the movie reruns that popped on the screen now and then, and I liked the classic movie *Heidi* in which she played the little girl named Heidi.

"Bill, I have no idea what she looks like now."

"Why's that so important?"

"Just one more thing I'd know."

This worried me, but I could see it was my problem, not his. I couldn't simply turn to the Internet and find enough information immediately. I'd have to visit the library for material on Shirley Temple, and I occasionally dropped by the Morton Grove Public Library, but I confess I was a bit lazy this way and figured I'd just wing it and maybe hope Mrs. Black still had the curls and a friendly face.

Mrs. Black became the US Ambassador to the Republic of Ghana early in the Ford Administration. By July 1976, President Gerald Ford appointed her as the first female chief of protocol of the United States, and I thought she could talk about her political life and her life as one of America's greatest child actresses of the 1930s.

I brought downtown the newsroom's Marantz recorder and a small microphone I owned that was not commonly used for field work because I didn't yet have access to the station's Electrovoice microphone. On the elevator, a very well-dressed man and a man with a camera with WMAQ printed on the side walked in first, and I figured they were probably going to the same place. We briefly exchanged names.

"Gene Randall."

"Hi, I'm Stew Cohen."

I looked at the cameraman. He just nodded and kept adjusting the lens on the camera.

I couldn't think of anything else to say. Well, I actually chose not to say anything else for fear of saying something stupid and exposing myself as a beginner at this type of work.

The elevator was far from ordinary. Besides looking very old, this was a private elevator with just one stop ... at the very suite where I was told Mrs. Black was staying. I knew the room number because I had jotted it down from talking on the phone to her contact. I glanced over at Randall, and he looked ready to greet anyone that opened the suite's door. I can't recall whether the door had a bell or he knocked on the door or even whether the door might have had some fancy security system. Then again I've probably confused technological advances that hadn't occurred yet to make security systems with cameras. Whatever Randall did to the door, suddenly someone was standing in front of us, staring back. She had curlers in her hair and a smile on her face. For a moment, Randall and I looked at her.

I formed words in my head that leaned toward, "Hello, could you please get Mrs. Black. I'm supposed to interview her this afternoon." I stared at a person I thought could be Shirley Temple, but I wasn't sure.

Randall stepped forward slightly in the elevator, "Mrs. Black, it's so good to see you."

My mouth opened! I'd nearly fired off innocent enough words, but words with a high level of embarrassment. The Richter scale's measurement for my embarrassing moment may have registered a significant trembler of stupidity on my part. I mean, she didn't appear ready to welcome guests, let alone WMAQ-TV for an interview. Maybe we were early or maybe Mrs. Black had been running behind schedule because of her commitments, but she clearly needed another half hour of primping time. I initially thought Mrs. Black had on an expensive-looking bathrobe, but whatever she wore, she remained a gracious hostess. She honestly seemed interested in her interviewers and perhaps more unusual, she didn't have a whole group of handlers in her suite, not that I could see.

Mrs. Black's reputation as a child star so many years earlier preceded her wherever she went. Shirley Temple was forever America's Sweetheart. Her smile appeared real, not an attempt to "put on a good face" for the reporters. She and I sat together in front of a coffee table around a huge pink couch that could have slept a couple dozen adults lying head to toe, one after another. She'd talk to the TV reporter first because he had to have the lights just so, and everything had to be in place, nice and neat. I sat nervously waiting. I tested my recorder several times as quietly as I could. I just couldn't sit completely still, so I pressed the buttons and watched the cassette move, and I turned on "record," and everything checked out. Mrs. Black occasionally looked at me and then again stared into the bright light, answering Randall's questions. Finally, they packed up their TV gear and got into the elevator. I stood and gave a halfhearted wave. All of this I could see, and she said good-bye, turned around, and now I was on. I say *on* in the sense the curtain rises, each knows his or her role, and the play begins. I take my seat again and start directing my actors—in this case just the one great actress. I lead her through every scene of her life that's important. These leading questions I'd formed in my head while I nearly disappeared on her huge couch.

"Stew, last time a reporter interviewed me, his microphone didn't work." I immediately decided not to test the interview on playback, not in front of her anyway.

"Would you like a glass of water?" Mrs. Black asked and waited for my response, but I could not answer her right away. I could not imagine her getting up from the couch and heading off into the kitchen of her penthouse hotel suite for a glass of water and then returning for her interview.

THE WYEN EXPERIENCE

"Sure, I'll take a glass of water," I said, like this was the most important glass of water in my life.

I sat, looking around the room, thinking I'll take her lead whether to be completely serious or joke around with her a bit.

"Here you go, Stew." Mrs. Black leaned over, setting the glass on a coaster, and then she sat down again. Her voice only briefly held the magic of her youthful self that America loved.

"Mrs. Black, tell me about the position you're starting as a representative for the U.S. government."

She launched into an explanation, and I have to admit, I wasn't so much interested in this as I was in her Hollywood life, but I'd get all the formalities done first. I don't remember what she said ... but Mrs. Black smiled. I recognized her political service was substantial, and she was to be taken seriously ... however, her smile gave the interview a pleasant feel. We got around finally to her childhood.

"Mrs. Black, what was it like as a child star, performing so often?"

"The dancing and singing were wonderful," she mused, and she told me about one of her favorite performers, Bill Bojangles Robinson. "He was such a great dancer. I remember him fondly."

She answered my questions with clarity and openness, and the play ended, and I turned off the recorder. We talked for ten minutes about the position she'd hold in government as the chief of protocol and how she thought that maybe she'd one day do something again in Hollywood, though she didn't seem overly enthusiastic at that moment.

The interview was over, and we sat smiling at each other. Mrs. Black made it easy. She asked me about what I did for WYEN and how I liked my job. I tried to sense how much time we could talk after the interview, so I looked for clues ... and then she stood up. I had already packed my recorder and microphone in my leather bag and just grabbed the strap. I headed back to the private elevator and turned around one last time, trying hard to burn the memory of Mrs. Black and the room into my head. I thanked her again and walked into the open elevator and hit the close button. I fished my recorder out of my bag and hit the play button ... and I heard her talking through the Marantz speaker. Then I hit the stop button and waited until the elevator opened at the hotel lobby.

I found my car and put the recorder on the passenger seat and turned it on again. I heard the same sentences ... but nothing beyond. What I found to my astonishment were a lot of blank spaces where I could swear her voice was supposed to say things like, "You've done such a good interview." I did hear strange beeping sounds though. Maybe I didn't rewind enough on the cassette tape. I did find a few seconds of her at the very beginning of the interview that I heard in the elevator, and although the only possible sound bite was far from the strongest statement she made, I took out my notebook and wrote a story and drove over to WMET. Although Dave Alpert and Bob Roberts were not in, I waited. Bob came back.

"This is the only sound bite," I said. "The rest of the tape is damaged."

He smiled, put out his hand, and I gave him the cassette. Bob copied the sound bite and gave the tape back. I knew he could work around anything.

The most unfortunate part of all this wasn't that I didn't have the opportunity for selecting the best sound bites from the interview, but that the interview was gone forever, and she said so much.

Back at WYEN, I sat with the story and then rewrote it, and wrote it again. The copy had better sound strong because the sound bite was weak. However, this being Shirley Temple, listeners wouldn't care what she said, just so they could hear her say something. Still, I faced the usual problem most radio broadcast journalists wrestle with in their early career. We haven't mastered our storytelling ability in thirty-second stories with a ten-second sound bite.

"Shirley Temple Black grew up from America's Sweetheart of 1930s musicals to today, working on a world's stage in the Gerald Ford Administration."

(Tape: 10 sec. xxx it's a great challenge "I appreciate an appointment by the administration. I'll do all I can as chief of protocol … it's a great challenge.")

"Many key lawmakers believe the love the world feels for Mrs. Black makes her a great choice. They recognize her new position helping foreign dignitaries in Washington DC also helps the president in international affairs. At the Ambassador West Hotel in Chicago, I'm Stew Cohen, WYEN Metro News."

This was my second feature story at WYEN. My first was with the great tennis player Bobby Riggs at Arlington Park (Arlington International Race Course) in Arlington Heights. Between the two, Riggs and Temple, there's one major difference.

The art of interviewing can always stand improvement at any stage of a broadcaster's career. Preinterview research and preparation, listening skills, concentration, and asking questions and followup are key areas of a successful interview. Those skills and techniques develop over time with plenty of interviews.

Sugar Daddy

My interview in the mid 1970s with Riggs, a great tennis player, points out a small amount of research, some listening skill and concentration, and a few decent questions and followup. My first feature interview held some promise, though I admit luck played the biggest part, thanks to Riggs's ability to talk with very little prompting.

Research and preparation are generally the easiest part of the interview. Why is Riggs at Arlington Park? What is his claim to fame? What makes him someone people want to hear about in my interview? Know something about him as a Wimbledon champion and about his Battle of the Sexes.

We already discussed listening skills in "The Body in the Backseat." Listening skills aside, the problem with lack of complete concentration is another area of importance. You may not be able to write notes as your guest is speaking because you are holding a microphone in one hand. However, you could talk for a while before recording and write the answers to a number of questions, and then do the interview and ask the questions again. The real interview may not have the spontaneity of the preinterview. However, your concentration will improve because you have a written outline or notes of what line of questions will work best.

The Riggs interview early in my career as noted in *The WYEN Experience* shows a man with plenty of confidence, energy, and creativity. Sugar Daddy showed how someone with extreme confidence and an abundance of energy can command excitement. I did not do a preinterview with Riggs because I knew he was completely media savvy.

The difference between the Riggs and Temple interviews is that I still have the complete interview of the great tennis player.

We met in a private room at Arlington Park Racetrack. The people behind the racecourse put on an annual boat show. They figured interest in Riggs could draw attention to the show. Riggs initially had his two beautiful models with him, but he asked both of the women to temporarily leave the room to give us privacy for the interview. I liked him right away. I didn't feel he'd have to put on a show for me alone. He'd been recognized as a showman, someone that always seemed to have a light shining on him at the front of a stage. In this private room, it was just the two of us and a microphone. The only lights were overhead, fairly diffused throughout the room. There was this slight echo because the ceiling was so high. Riggs spoke in rapid, staccato tones. I spoke slowly, not completely sure of the direction I was heading with my questions, but I pushed through, thanks to preparation.

Stew Cohen (S.C.). I'd like to ask you your philosophy on life. What is your philosophy on life?

Bobby Riggs (B.R.). What was that question, Stew?

S.C. What is your philosophy on life? How do you look at life?

B.R. (First two sentences are said by Riggs in a sing-song poetic manner.) My philosophy, the way I look at it, is to try and be happy and have a good time. Go places and see new faces, and try to stay young in heart, be of good cheer, and always try to see the brighter side of life. I try to walk on the sunny side of the street, as the saying goes. Never take yourself too seriously, get a lot of exercise, eat good healthy foods, get a lot of sleep, but exercise, have a long life and a happy life. Because if you get lots of exercise, you should be in pretty good shape, and you'll live to be long, and my ambition is to play tennis when I'm ninety years old just like the king of Sweden did. He played up to ninety.

S.C. Do you plan to return to Wimbledon to play?

B.R. Yes I do. We're having an old-timer's tournament this summer, starting June 20 at Wimbledon this year along with the regular tournament. It'll be all the past champions in singles and doubles that want to play. Sixteen pairs, a special tournament. So I'm going to try and get somebody like Pancho Gonzales or Dick Savitt or Vic Seixas or one of those past Wimbledon champions who's still playing and team up with them and try to win that special tournament, even putting prize money up for it for the old timers.

S.C. You played Dr. Renee Richards in the past when she was a man. Now she's a woman, and you want to play her?

B.R. That's right because we could have a big television special, and the whole country would like to see how Renee Richards could do against Bobby Riggs because she's wanted to play in the top tennis tournaments for women, like Forest Hills, the Australian Open, Wimbledon, and so far, the girls have not let her in the big tournaments. She has played in several small women's tournaments, but not against the big stars, so it looks like the girls will not let her play in the big tournaments. I've played two great mixed-sex matches—one against Margaret Court, one against Billy Jean King—so I have a track record. I am one and one with the two best girls in the world, so if Renee Richards can't play the girls, she should play Bobby Riggs because Bobby Riggs is 50/50 with the girls. She would know how she would do against women because if she beats me, she would be one of the top women players in the world, and until women let her in the tournament, she's not going to know any other way. So I'm going to be a good sport, good guy, and play with Dr. Renee Richards.

S.C. Does she know about that? Have you both talked to each other about playing?

B.R. Yes, we've been talking about it, both want to play. We're negotiating now with the networks. We may do it for CBS, may do it for NBC or ABC, maybe we'll do it for the network putting up the most money. Who knows? We'll play for the best offer, and we hope we do it within the next six months. I hope you come out and see it.

S.C. Will you definitely play?

B.R. I'm looking forward to it. I'll give her some chops and lobs and slices, and I hope I can psyche her out just like I did Margaret Court.

S.C. I'm not sure about this, but did you have a dog sled down Michigan Avenue?

B.R. That's right! When I first came to Chicago Tuesday morning down Michigan Avenue where the Lake Shore Bank on the corner of Ohio and Michigan, I went down Michigan Avenue. Then we turned to the right at McClurg Sports Center, and I had a great team of dogs—four big Malamutes up from Minnesota way—and they were well trained. I had some of my Nabisco Milk Bone Biscuits, and they loved them, and they didn't give me any trouble, and they were just a beautiful team. That's quite an experience. How many guys do you know that have driven a dog sled team down Michigan Avenue?

S.C. You're planning to play Dr. Renee Richards. What else do you have in mind to do?

B.R. Well, I have a big bicycle bet coming up. Evel Knievel bet me I wouldn't ride a bicycle from California all the way to New York as soon as he puts the money up. One-hundred thousand dollars. I'll strike that, probably be in August. I might go to China in March because the fellow I beat last year in the Death Valley Run, Bill Emmerton, wants a return match this year. He said, "This time I'm taking you to the Wall of China, only two thousand miles long." I don't think we'll run the whole two thousand miles; I think we'll just run fifty-mile sections, probably a good three to four sections, and after the run is over, I'll challenge those Red Chinese table tennis players. I'll play some seniors over there, some of their better senior table tennis players. I know it'll be a great experience there. They say the wall is uphill all the way!

S.C. Thank you very much.

B.R. Okay, Stew Cohen, you are terrific!

We walked out of the interview room.
"I hope you play Ping Pong," Riggs casually tossed a challenge my way.
"Sure, I've played."
He led me to a table in the next room. Riggs had already psyched me out, but I figured my years of playing Ping Pong with my high school classmates Monty Abrams and Ian Wolinsky might pay off in a minimum of embarrassing moments in front of several hundred people.

This was a little different from the time in college where I got to shoot pool with Rudolf Wanderone, "Minnesota Fats," and wrote a story on him for WIDB Radio. Wanderone and I didn't play in front of several hundred people because I had scheduled our billiards interview before the crowd came, but at Arlington Park, hundreds of people were seated around a Ping Pong table waiting for Sugar Daddy to emerge from our interview and play people in the crowd and then talk about exercise and his exploits. Riggs gave me a paddle and introduced me, and I noticed the net was a bit higher on the table than I was used to on my table back home. I think the Riggs net was regulation height. I slammed several times, but the ball kept going into the net. Riggs led the entire way and hit everything back. A few balls bounced on the table and hit me. A guy got up from the audience before our game was over and approached the Ping Pong table. He looked bored watching me and offered to play Riggs, hinting he could give the tennis champ more of a challenge. I found that kind of embarrassing, but I did not think for a moment I'd win. The game ended, and Riggs thanked me, and I thanked him. Then I picked up my recording equipment from under the table, and I left the building.

Riggs died on October 25, 1995 in California at the age of seventy-seven.

Swims Like a Fish

Just as Riggs's vocal display rocketed serves, smashed forehand drives, and lobbed sky-high backcourt gems in his use of the language, Mark Spitz and I waded into the depths of his oral pool.

On my way to the Olympic Torch, a Mount Prospect sports store where the great swimmer agreed to talk to me about his new line of swimwear, I thought about what the two of us had in common, and by a process of elimination (he was bright, athletic, had model good looks, and liked water a lot), I determined with certainty we both liked water a lot. What could I ask Spitz? If knowledge were liquid, his understanding of swimming could fill an Olympic-size pool while mine might not fill a kiddie pool. My experience in the water couldn't even produce a ripple because I didn't dive and never swam competitively.

Spitz was the talk of the 1972 Munich Olympics until terrorists killed Israeli Olympians. The horrific images of the terrorists watched by a worldwide audience completely changed what was the feel-good story of Spitz winning seven gold medals.

I had a few minutes before Spitz was ready to speak to me in the store. I closed my eyes and brought the pool back in my head. Spitz became visible, swimming again, leading all the other great swimmers from a half-dozen countries. With his mustache a sign of defiance for those believing a clean-shaven swimmer could go faster, Spitz skimmed through the water almost as fast as a speed boat. Spitz's arms were a blur, creating nearly a circle at his head, virtually an outboard motor churning through the water.

Suddenly my mind switched from Spitz raising an arm, the universal recognition of Olympic glory to ...

"Watch, Mommy. Watch me!"

On a hot August day, I waded into the cool, clear pool at Nippersink. I hoped my mother would cheer my ambitious attempt at swimming one width in the resort's pool. Far cry from the millions of people cheering Spitz for every one of his Olympic heats.

"Watch me swim," I commanded.

She put down her book and leaned forward in her lounge chair, accidentally pushing my inner tube with her foot. The sun had gone behind a cloud.

I stretched one arm beyond my head, hunched over the water, my mouth inches off the surface, and pushed my other arm back like I learned in summer camp at YMCA Leaning Tower in Niles. I twisted my head under my armpit and looked again at Mom. She watched me splashing my way through the water, stroking with my right arm over my head and back into the water, and doggy paddling with my left arm.

"You can swim like a fish!" Pride gushed from Mom.

I barely swam the width of the pool and could have easily filled a lobster tank with the water I swallowed! Having her watch my attempt at swimming got me jumping up and down in the shallow end of the pool, but my little show was plenty. The sun came back out. She looked up briefly and then reclined, positioning herself for the best angle at the sun. I was done showing off. Now I'd just hold my breath underwater.

Concentrate for a new underwater "holding your breath" record, but I surfaced within a few seconds, and Mom interrupted a second attempt.

"Sure you don't need the inner tube?"

"Watch me one more time!"

She really wanted me out of the water, maybe for more suntan lotion or possibly to stop me from interrupting her sunning.

"I don't need the inner tube. You just said I swim like a fish."

She sat a little straighter in her chair.

"Okay," I reluctantly surrendered to her stubbornness for my safety. I walked up the steps in the pool and reached under her chair, grabbed the inner tube, rushed back to the pool's steps, and squeezed into the tube. Mom watched, relieved. She didn't see me tossing the tube to the side of the pool. Although I might not have been a "swim-the-length-of-the-pool swimmer" and I couldn't put my head in the water, I was still a proud "swim-the-width-of-the-pool swimmer with my head out of the water and no inner tube in sight." Maybe that's how Spitz started?

A Spitz interview drew my attention because he'd been an icon and could fill in blanks on what his great feat of Olympic measure meant in the backdrop of the Munich Massacre.

"Mr. Spitz, you've been the greatest swimmer we've ever seen. All those Olympic medals, but then terrorists struck. What do you remember about that?" His exact words in the interview were not saved on tape, but my journal held how I interpreted what he had said. The horror of what the terrorists did showed in the sadness in his eyes. He wasn't allowed to stay in Olympic Village.

"They rushed me out and said very little. However, I'll never forget the horror of what the terrorists did." The enormity of the Munich Massacre weighed heavily on Spitz, who told the media that he and many others could only think about the people whose lives were taken.

His Olympic medals and his line of swimwear just didn't carry the weight of such a time of great tragedy, but I asked about the swimwear anyway.

"This is one of the stores carrying my line of swimwear."

"Can someone swim faster wearing the trunks?"

"Maybe a little faster I think, but probably not compete for the Olympics."

Perfection motivated Spitz to swim faster, train harder, dedicate himself to becoming the greatest in his sport. He didn't have to spell it out for me. I could understand the expressions on his face because he was mostly subdued in my interview.

I had no trouble recognizing Spitz inside the Olympic Torch. He'd been in the back area of the store. He hadn't changed much since his Olympic success five years earlier. Physically, the Olympic champ looked in great shape. He was friendly and willing to answer my questions, though he wouldn't venture anything beyond what I asked. Yet I didn't expect Spitz to sound or act like Riggs in our Sugar Daddy interview. Had I been better at the interview process, I probably could have opened him up, but this was only my first full year of professional radio. I just wasn't aggressive enough or probing enough. I thanked him for his time and truly meant it because for a while, I found greatness staring back at me.

Mr. C.

For the Tom Bosley interview, my problem wasn't so much the phone, but the fact his voice was so recognizable, I had a difficult time focusing on what he was saying. I kept thinking I was talking to Mr. C. of the popular television show *Happy Days*.

The show brought us middle-class America of the late 1950s and 1960s and a father many of us wanted our dad to be like. This would have been a good theme for an interview with Bosley on Father's Day, but I'd have accepted any opportunity to talk to Mr. Cunningham. Bosley, I was told before the interview, was coming back home from Hollywood to the Chicago area for an event in 1977, and WYEN would mention this in our interview, and I had Mr. Bosley on the phone.

"When we started out (*Happy Days*), the nostalgia trip was very strong in the public's mind."

Bosley hinted at the fact we longed for a more happy time when life did not seem as complicated.

"You return every year to the Chicago area?"

"I love the Windy City. A lot of my family still lives there. I'm a dyed in the wool Chicago Bears and Cubs fan."

I know this is crazy, but for a moment on the phone, I imagined I was really talking to Mr. C., and we were on *Happy Days*, and he told me on the phone to do great at my job. Fatherly advice, maybe, but fantasy works wonders with the actual actor on the other end of the line.

What I liked about Bosley's Mr. C. was his voice—so soothing and rich and deep, and it would make you relax and lower your stress like drinking cherry-flavored Pepto-Bismol after a stomach ache. I wish I could have bottled his voice.

On October 19, 2010, Bosley passed away at the age of eighty-three in his California home. His television family, his real life family, and millions of fans mourned his death. I thought about his passing and how our paths came together for a moment and how I can still hear his voice in my head, and it's soothing and quieting.

I can still hear a dark-haired woman skillfully producing sounds from her accordion as she modulates the higher and lower register of her voice.

The Love Goddess

How many guys can say they had a date with "a love goddess?" This is not Greek mythology here, but real-life stuff. Some of you have known her as "Aphrodite of the Accordion." Others recognize her as "The Love Goddess." I knew her briefly. To me, she was a comedian playing an accordion, but also a free spirit. Her dark hair and very white skin made her stand out on stage, back stage, anywhere she chose to be seen and heard.

The Comedy Cottage in Rosemont had a stage for its comedians and many tables for customers to sit around the stage. I was looking to do a feature story for WYEN and was told the Comedy Cottage housed a stable of very funny comedians on the rise. The restaurant was less than a mile away from WYEN, so this seemed natural for the WYEN audience to learn about a place where they could go for laughs, a beer, and a good meal. The owner at the Comedy Cottage got me together with the comedy emcee, Larry McManus. Larry was just a regular guy. He didn't have "his on switch on constantly," and I liked that. He offered a free meal, recommended the best hamburgers in Rosemont, and brought several comedians to me before they went on stage.

I returned a week later on a performance night and sat down at a corner table with Larry. I turned on my recorder and interviewed him about the history of the Comedy Cottage venue.

"Comedy Cottage offers comedians a place to develop their craft, and we even have an open-mic night for people in the restaurant to perform on the spot." He grinned. "Sometimes they are really funny because they're talking about themselves, an area they know well. But I've also seen some bomb, even the performers that had been on stage before." Larry shook his head, still grinning. "We had one regular have such a bad night he started swearing and stormed off stage."

When I got to Comedy Cottage, the place had maybe a dozen customers, but as Larry continued talking, I noticed every table was now seated with diners, looking for a night of great comedy.

"Stew, I think you should get up on stage. You'd make a pretty good standup."

"Larry," I nearly interrupted, "I don't think I could go up there, but thanks for the confidence."

Besides Larry, I remember only one of the three comedians I interviewed that night. The first two comedians were men, and I'm sure they said some funny things in my interview, but only the third comedian stood out for her unique comedic approach. She brought an accordion to our table. She asked if she could put it on the table. "No problem," I answered. I wasn't sure whether she was a musician or a comedian, but she had this mesmerizing presence in voice and physical appearance. She told me about her family, why she performs, and gave a preview of some of the comedy material, but I unfortunately can't remember what she said, and I didn't keep the interview. She thanked me for the interview and left for her turn on stage. Larry introduced her. I sat with my burger and beer, ready for a few laughs.

Larry wasn't super funny, but I could tell he wasn't trying to overshadow the talent; his was more of a mix between comedy and the business of moving the show along. Finally he came to her, the accordion lady, held for last, I guess because he felt she was what many were waiting to hear. She'd say something in a very loud voice and then stare at the audience for a moment, blink her long eyelashes together, and then play a few notes on her accordion. Men as slaves came up in her routine and other unique bits that made her different from anyone else I'd ever heard.

I told Larry I'd put together the feature story on the performers, air the story, dub copies, and return to Comedy Cottage.

I returned to Comedy Cottage and gave a tape to Larry. Instead of giving the other tape to her, I asked her to join me at a radio station party, and she agreed to go. Garry Meier was there at the party, so were other WYEN announcers, but I can only remember Garry sitting on a chair greeting my guest. Seemed my "date" was the hit of the party, and she held her own very easily. We couldn't stay particularly long because she had to perform on stage that night at Comedy Cottage, so we got in my car and headed back to the restaurant. We parked in the restaurant lot, and I played the cassette tape for her, or tried to play it. Something wasn't working right, and I pulled the cassette out and found the tape had been winding around the pinch roller and would need serious readjustment, but I didn't want the embarrassment of this, so I told her I'd get a different cassette to her later. Always an awkward moment, and this was the moment. I mean, this wasn't my definition of a date; more of a continuation of our interview. We didn't spend the entire evening together, and I didn't hide all my character flaws, and I didn't really learn that much about her, nor did I think she was thinking this was special.

Maybe I was over thinking, but that's what I thought in the minute or so it took me to make up my mind, skip any kind of affectionate response, and rush out of the car to open her door. As we walked from the lot to the door of the Comedy Cottage, not only could I see my breath, but I could see my thoughts of possibly getting together again disappear in the cold air as did my breath. Inside the restaurant, for just a moment, I convinced myself to ask her to lunch or dinner, but she had already disappeared into a group of comedian friends standing in a circle near the restaurant's entrance. I couldn't redirect her attention without directing the attention of all those funny people, and I knew I didn't have a chance with them now paying attention to me. So, I turned around, faced the night, a slave of my own shortcomings, and said good-bye in the cold to Judy Tenuta, and although she'd never hear it, I wished her great success as "The Love Goddess."

I sent this story to Tenuta's agent, and in turn, Judy read the story and commented, "This brings back many happy memories."

Judy has provided many happy moments for her fans since she performed standup comedy at the Comedy College in the 1970s. Her body of work has taken her from a variety of comedy themes and venues, such as *Hollywood Squares* and other game shows, to writing the book, *The Power of Judyism*, and producing comedy records and lending her voice to animated programs and film.

Working on the Tenuta story offered a good example of human interest, information in the area of soft news. Some radio reporters are skilled in both soft and hard news, making them valuable for covering a wide range of stories. Not unlike the Tenuta story in its human interest aspect, the next story is pure hard news with an historic leaning.

Tokyo Rose

Stories with an historic tie-in were most intriguing at WYEN, as they continue to be in radio news coverage. Being "in" on history has an appeal because it's that step beyond reading about historical events. Most of what reporters cover is the everyday type of event. This story is an exception.

My father, Sid, and my grandfather, Max, provided a solid foundation in which to approach this story. My dad and my grandpa served the United States in the world wars. Grandpa Max was a horse soldier in World War One, and my father was an infantryman in General George Patton's army in World War Two. Max rode a Cavalry mount, and Sid slogged through one of Normandy's beaches, carrying a Browning Automatic Rifle BAR on D-day. Grandpa resupplied bullets and sometimes supplied new rifles to the doughboys, and Dad fought Nazism. Reading about the wars in high school history class always fell short of the fact I had two living witnesses to the fight for protecting freedom, and although they didn't tell me harrowing war stories, they did let me in on enough of the pain they suffered to understand the effects of war.

My news director at WYEN asked me to cover Iva Ikuko Toguri D'Aquino's news conference. He didn't ask me by invoking D' Aquino's name. I didn't know who D'Aquino was in world history, but then Wilkinson put aside her legal name and directed me to cover the woman the GIs had dubbed as Tokyo Rose, broadcasting propaganda on Radio Tokyo. President Ford had pardoned Toguri unconditionally on his last day in office in January 1977. Bill didn't have to wait long for my answer.

"Where do I go for this?"

"Downtown." Wilkinson already had an address scribbled on a piece of paper: Belmont Avenue just off Clark Street in the Lakeview neighborhood of Chicago.

"When?"

Bill wrote the date and time on the piece of paper, and I put it in my pocket.

I accepted the overwhelming view on Tokyo Rose—a traitor during World War Two. I hadn't realized *Tokyo Rose* was the name given to all the women that were hosting American music and talking directly to our soldiers on Japanese radio during the war. Some said things demoralizing for our troops.

Mr. Ford did not side with the strong notion of Toguri as a traitor, at least not this Tokyo Rose. He'd heard details from media investigative work on how two key witnesses in a 1949 trial were coerced into saying Toguri was broadcasting propaganda, when in reality, they told a *Chicago Tribune* reporter twenty-seven years later that she had not said anything against the US military. Ford's pardon got the immediate attention of people that lived through the war, but by the time of the *Chicago Tribune* series, many people were already well into their adult years with children that really didn't have a connection to the war. Still, the news conference for Toguri's reiteration of her story was something worth covering. Nothing better than a compelling story for radio.

Sometimes in covering stories, a reporter confronts his or her biases. These biases can taint a story, and that's the reason the best reporters search for balance in their interviews. It's hard enough covering stories with differing viewpoints without giving more weight to one side or the other. You might subconsciously pick a stronger argument for the side you believe and use it in your story. Attaching our own feelings only complicates what is already layered with subjective interpretation. We could easily leave out pertinent information, stress parts of the story to support our bias, decide not to seek out comments from another viewpoint, or do things to confirm our preconceptions. This story of Toguri's pardon was undoubtedly a real challenge because I carried some editorial baggage, believing she was a traitor from what I'd been told for so long a time.

Toguri was an American of Japanese descent stuck in Japan after the United States declared war on the Empire. The bombing of Pearl Harbor sealed her fate. I had thought she had spoken out against our soldiers, but those investigating her case believed she had not demoralized our troops.

J. Toguri Mercantile, where the news conference had started, drew reporters from all over Chicagoland and network reporters. I was late. Parking spots near the store were taken with television trucks. I found an open door and rushed into the front of the store. However, Toguri was in the back of the store for the news conference. She stood so far away, I could barely see her. I caught an image in the seconds I saw her between a couple of cameras, and it appeared she'd been sitting at a table, eyes blinking from the strain of the bright lights, watching some government guy in a suit reading the words of President Ford's pardon. I thought she wore a white shirt and black pants, but the lights made it hard to see the color of her clothes or whether she was trying to smile.

Reporters inched closer. Toguri felt them. With their words in years past, they had caused her to lose any privacy she'd sought. The pain didn't show, despite the camera crews refocusing on her face for a tight shot. She waited her turn. Others finished reacting to the pardon. None of what they said mattered, and although the camera lights were on, only a few camera operators seemed interested. And then she stirred; Toguri rose, her posture straightening. Such a rush to the cameras, the operators woke up the room. Someone pushed a microphone on a stand on the table where she stood.

All those forced broadcasts more than thirty years ago were not lost on her. She didn't seem overly nervous despite the obvious duress one might feel under the light of a microscope. Now, in this store, filled from front to back with reporters, she'd string real words together from her heart, rather than words written by someone else in a script. The threat of death no longer whispered in her ears. Ford's belief in Toguri freed her. She could finally express herself without fear of incarceration. The words had waited inside her, overlapping each other, sentences crowded into each other, layer after layer for all the years of staying silent, afraid to stir anger that lingered in America from the war. Until now, no one really heard her. It was ironic that she couldn't be heard in the years after her broadcasts to our troops because the ramblings of Tokyo Rose were the very essence of a traitor's message ... but not Toguri's message. I got stuck in the back of the store, eight or nine rows deep with reporters. I couldn't hear much, so I had more of an opportunity for philosophical thought.

The price of freedom blended evenly into a fluid story of my grandfather escaping from a small village in Russia to a bustling Chicago. Waiting for words of Toguri's personal experiences, I recalled the fight for freedom that pulsed in the very blood of my grandfather in Pinsk during the beginning of the Bolshevik Revolution. At the age of sixteen, my grandfather paid into the price of freedom by leaving his parents behind at the urging of his father, David. Max remembered his father's words—among the last words he'd recall from his home because he repeated those words fleeing as he did with his thirteen-year-old brother, Morris.

"The only life worth living is a life away from the control of the Czar."

Their mother's tears moistened one day's meal. She gave them what she could, and Max and Morris carried the food away from the family cottage. With no certainty they'd complete all the legs in a mass exodus of their people, they ventured forward to the United Kingdom and then to America on a ship. My grandfather and uncle, then just teenagers, not only escaped the Bolshevik Revolution, but arrived safely at Ellis Island, completing a very long journey. Max carried a secret with him, the price for freedom, and soon he'd have to deliver on the secret he kept from Morris. Max agreed to serve in the military as a mere teenager and signed conscription papers, sealing the promise.

The real test of freedom asks for sacrifice, for without sacrifice, what truly are you willing to do to fight for freedom? Toguri said she never went along with the program forced upon her in the war broadcasts. She said her broadcasts were never anti-American, not something one might construe coming from the mouth of a traitor. While I thought about her sacrifices and how the message got so scrambled, I thought too about how much my family was prepared to give up. That's when the image of Max on a horse struck me. He put his life in harm's way so his younger brother might live in freedom. His sacrifices weren't some verbal display; it was the reality of Max's soldiering as bullets fired by the enemy narrowly missed him but struck two Calvary soldiers on either side of his horse. The spot they fell was the spot they died. This reality changes one's soul, and for one of the last cavalry soldiers in war, Max stamped his fight for freedom in the Battle of the Argonne Forest.

"When you live a part of your life without freedom," he carefully explained, "you recognize just how precious and vulnerable freedom is and how you must protect it always." For his faith in freedom, Max suffered shell shock the rest of his life, but he willingly and quietly suffered.

My father's story measures freedom by the distance he traveled with the 137th Infantry into France and Germany despite suffering from spinal meningitis and trench feet. He had a chance to leave the army after he contracted spinal meningitis, but my father believed his life was rightfully in the hands of General Patton and Company F. He waded ashore on Omaha Beach in Normandy, the twelfth wave on D-day. The distance he traveled and the sacrifices he made for freedom were recognized in the form of a Purple Heart.

January 19, 1977 wasn't so far removed from the end of World War Two, only thirty-two years, and a lot of veterans were still alive and wondering why Tokyo Rose would receive a pardon. I'd want that story for my WYEN audience, but I'd have to be willing to listen and not take my grandfather and father's sacrifices with me.

I could not squeeze my way up front to improve the quality of the recording; the news conference was too far along. I tried inching my way forward, stretching the microphone as far as I could, but I didn't realize the cord came partially out of the recorder until after Toguri thanked President Ford for finally freeing her of the traitor label.

"This is an act of vindication," Toguri stated, preparing to try to block out all the injustices hurled against her for so long.

Ah, but could I free her of the label? Could other reporters do the same? The history I brought with me of military sacrifices by my grandfather and father would probably make one think I'd editorialize this story.

This was a frigid and wet January day, and many reporters wore all kinds of winter wear. Someone walking by, looking in Toguri's window, might think the store was as cold as the outside was. Reporters kept on their coats and boots, mostly because there wasn't any place to put them, and the floor was extremely wet and slippery. I looked on anxiously for an opening in a ragtag army of reporters. Occasionally, a glimpse here or there of her face, an arm, or her shirt teased me with the thought I'd see more, but a wall-to-wall phalanx of cameras made that hope unrealistic at best.

Evidence and time provided a new look in 1977 at only the seventh person to be convicted of treason in the United States. I don't know whether I was thoroughly convinced of her innocence as I stood there with my microphone, but I recognized that for WYEN's audience to hear an unbiased story, I'd have to shake loose of these preconceived notions on that day and report on how she'd finally been freed with a presidential pardon. She was finally a free woman, no longer tied to Tokyo Rose—just Iva Toguri, and that was all she ever wanted. At the age of ninety, Toguri passed away in Chicago. The year was 2006.

A World of Politics

Chicago Mayor Richard J. Daley earned an unshakable loyalty from an army of city workers. Upon his death, I sat in the Chicago City Hall newsroom, waiting for former alderman and then Acting Mayor Michael Bilandic to assure us the city wasn't going to fall apart without Daley at the helm. Bilandic addressed the media, but he lacked the Daley charisma and didn't know how to work smoothly with the media. He pulled in one direction, the media in the other, and so everything Bilandic did was examined as though his competence on the job was questionable. In the light of misgivings about the new mayor's administration, we sat and waited for something unusual to pop up at the podium. What popped up came from the row in front of me, and he looked like the actor Joe Don Baker from the movie *Walking Tall*.

"Hi, I'm Mort Crim," he said after turning around to face me.

His baritone broke through the silence in the room ... silence out of reverence? I think so. Reporters in this room, except for Crim, covered City Hall under Daley. They knew about "Hizzoner, the Mayor." They felt a kinship and a significant loss.

Crim's salutation floated above our heads, breaking an obligation for silence, but I also felt obligated to engage Crim. Yet I wondered, *Why me?* I never asked a question at City Hall or talked to the veteran broadcasters and journalists. I hadn't yet earned the right to talk to them or ask a question. I was pretty much invisible out of respect for a rookie code I felt obligated to follow, but truly more so from self-paralyzing introversion. Crim waited for me to say something. He wasn't turning back to face the podium. I had watched him anchor a WBBM Channel 2 newscast the evening before. I thought his friendliness at City Hall was good—but misplaced with me. Crim smiled, dressed in his fine-fitting suit. His teeth were ivory-soap white. Here he was smiling at me with the actor's beefy profile.

"Hi, I'm Stew Cohen of WYEN Radio. Congratulations on the WBBM TV anchor post."

I didn't know whether to address him as Mr. Crim or Mort, so I didn't address him. I said no more because anything else and I'd have sounded disrespectful of the quiet. Bilandic walked into the press room from his city hall office attached to the media room. He had an air of arrogance on his face that reporters say they saw too. I covered him this one time and saw a mixture of arrogance and feistiness. Eyes wide open and a half smile came across as arrogant but may have been unintentional on his part. In a closed-door negotiating session, the Chicago City Council had chosen the Bridgeport alderman for a six-month term as acting mayor upon Daley's death in December 1976. Crim sat back in his chair, watching. I expected a long answer from Bilandic. Would he run for a full term? Bilandic slid directly behind the podium, set upon immediately by reporters. I don't know whether Harry Golden Jr. raised his hand, but the first question came in his general direction, and Bilandic did not ponder it. Would Bilandic run in the special election for filling the final two years of Daley's term? I was preparing to be confused with obfuscation. He'd talk and talk, but we'd not learn a single thing. Surprise!

"I'll run," Bilandic assured, smiling at us as though we already knew his intentions. Reporters' excitement was clearly evident. They moved forward in their chairs, pens scribbling on paper, sensing a good front-page story of Bilandic wrestling over how he came to this decision. Then the hot air balloon of a story ran out of air and plummeted to the ground.

"Just head over to Grant Park tomorrow morning, and you'll see me run."

Front page ... make it back page!

This was Bilandic at his media-baiting best. Feisty, funny, and arrogant; he was all of this. Bilandic served as mayor from 1976 to 1979, and then he lost to Jayne Byrne. A paralyzing winter storm and its aftermath of snow-covered side streets angered Chicago voters, making them turn away from the machine, and they voted for Byrne. Bilandic had underestimated the effects of the storm on voters. However, he went on to become chief justice of the Illinois Supreme Court and passed away at the age of seventy-eight in 2002.

In five decades of covering politics and the people representing us in Washington DC, Illinois, and local government, I found Bilandic an exception. Most I've reported on have worked fairly hard on finding out something about the reporters covering them on a regular basis. Congressman Abner Mikva held a *Fun Run* every year he served as a US representative for the Tenth District in the northern suburbs. He started the runners and welcomed them at the finish at Harms Woods, a few miles from Old Orchard Shopping Center. That's where I met Mikva for the first time. I quickly learned congressmen were typically very friendly, maybe because they'd have to run for reelection every couple of years. Mikva knew me by name and treated me with the respect I'd assume a veteran journalist might receive. Bob McClory treated me like a son, Phil Crane treated me like a lunch companion, and Don Manzullo treated me like a son-in-law. I didn't meet with Dennis Hastert often enough. Governors were harder to contact for comments. I'd have to cover an event where they spoke. Interviews with Illinois Governors Jim Thompson, Jim Edgar, George Ryan, and Pat Quinn were special occasions. On one of my favorite interviews for WYEN, I talked to Governor Thompson outside a meeting room of people waiting for him. I knew he wasn't in the room, so I waited near the entrance and asked him for a specific comment on an issue, which generally worked well. Thompson's ad lib ability rivaled the best orators. He thought on his feet as well as I'd ever seen. Of the governors, I was probably closest to Pat Quinn because of his early media attention as a political activist working to amend the 1970 Illinois Constitution and as head of the Coalition for Political Honesty to shrink the size of the Illinois House. I'd call him every couple of weeks for a quote on his latest efforts, and he was always accommodating.

Chapter Fifteen

PROGRAM DIRECTORS

Wayne Allen

"MacArthur's Park is melting in the dark." Donna Summer's voice echoed down the hallway … "All the sweet green icing flowing down …" Allen set his watch for seven minutes. On the turntable, the full version of the 1978 disco song gave WYEN's morning announcer enough time for a bathroom break. Wayne left the studio, rushed down the hall, did his business, and returned to the studio before Summer started winding down, "… and I'll never have that recipe again." Every announcer needs a bathroom song, and this was his.

Wayne, whose real name is Alan Weintraub, was the no-nonsense, often well-dressed program director of WYEN. He'd been PD in the mid to late 1970s, right through the disco years. Allen was always aware that his on-air staff had offered listeners excellent music hosted by talented announcers. He fronted the lineup in the morning, but the deejays were strong throughout. Meier followed Allen for midmorning, and Reynolds and Zur carried WYEN through into the evening before Mike Roberts stepped in to bring an overnight loyal listenership along with him. Other quality broadcasters entertained for various periods of time, including Val Stouffer, Dan Diamond, and Mike Tanner.

Allen had every right to treat WYEN as a training ground for young talent. For some, this was their first radio station; others listed WYEN as their second station. Either way, WYEN came early in the resume of many on-air announcers, leading me to ask Wayne whether he too viewed WYEN as a training ground. The answer he gave made an impression and led me to consider a revision of my initial thoughts as I began writing *The WYEN Experience*. He viewed WYEN as a solid competitor to Chicago radio. The talent he supervised at the suburban radio station with a 50-thousand watt signal always strove to do the best job possible. Wayne explained he would find their strength and have them use it to benefit listeners, and he noted the staff never appeared to think they were in training. Allen recognized each was already well versed in music and entertainment.

Could this great format work again? I wondered. Wayne quickly answered, possibly having thought this out before.

"In today's culture, Request Radio would work far better than in the 1970s. Everything is easily assessable because of the Internet. You'd have a couple of HD channels, digital equipment, cell phones, and would be far more successful today than yesterday."

Wayne's yesterday consisted of a hot clock for programming by hand, a razor blade, editing block, turntables, cassette and cart machines, and a reel-to-reel recorder.

"Everything we did was manual, (laughs) … manual labor as in marking work tape with a grease pencil, splicing two ends of tape, then playing back to make sure the edit did not affect the new sound."

Allen's show had elements that weren't always in the Ed Walters playbook on radio.

"I did character impersonations between 5 a.m. and 6 a.m. These were drop-ins, unfortunately not approved by Mr. Walters, but they were funny, such as Tanya Turkey for Thanksgiving or Christmas Time Santa Claus as Sam Claus."

For twelve years, Wayne was an executive with Hair Performers Corporate, working as senior vice president of marketing and advertising. Today, he is senior account executive for Navteq Media Solutions. Between WYEN and Navteq, he'd held high-level positions in the marketing and advertising industry, such as marketing sales manager for WLS-TV Chicago in the mid to late 1990s and president of Alan Weintraub & Associates, freelance consultation associated with media and the entertainment industry.

Allen, Meier, and Reynolds were most closely tied to the disco years through the Savoy in Wheeling. Wayne managed the bar and said Garry worked for him as a club deejay on weekends at the disco. Allen found irony in his connection to the Savoy and disco music. He related to the irony of Disco Demolition Night in the summer of 1979.

A Jerry Westerfield shot of Allen for publicity photos.

"Ironically, when I had to let Garry go, it was only two years later that he and Steve Dahl went over to Comiskey Park and blew up disco records."

Allen's time at WYEN ended one evening in 1979 after a stint of five years. Wayne recalled Mr. Walters called him into his office during the 6 p.m. newscast.

"He told me he had to let me go … said I was done."

Although it went unsaid, Allen believed his departure was tied into his request for a raise.

"My wife didn't know where I was because I never came back into the studio to follow the evening newscast."

Would Allen do a WYEN show all over again if he had the chance to return to the 1970s as a young man in his twenties?

"Yes, I'd do it again if I could go to work in a T-shirt and jeans."

―――――――――― THE WYEN EXPERIENCE ――――――――――

Jerry Mason

"I spent a lot of time during lunch hour in my car monitoring our direct competition."

This *road warrior* life led to an unintended consequence. Some staff members speculated that Mason, the WYEN program director, wasn't just driving around monitoring other stations. Jerry did not know about the rumors during his time at the station from March 1979 to February 1980. He now laughs about what he calls "the ridiculous nature" of the speculation.

"The rumors were of course false that I was doing cocaine every time I went to lunch," Jerry said.

In our phone conversation from his home in Sibley, Iowa, Mason said he did not hear about the cocaine rumors until he left the station to form his own "consultancy" called the Jerry Mason Group. After Wayne left WYEN, Walters hired Program Consultant Paul Christy. He worked at WYEN for a short time until Walters hired Mason and put him on the morning show, 6 a.m. to 10 a.m. During Jerry's time as host of the morning show and program director, he had a lineup of Louie Parrott working midmorning, followed by Roger Leyden (Wayne Bryman), and Mike Roberts on overnight.

Parrott contributed stories to the Walt/West Enterprises *Outside the Loop* monthly magazine for Chicagoland. Inside the May 1979 issue of *OTL* was an article Louie wrote called "The a.m. Sound on WYEN is Positively Jerry Mason."

There's a new morning sound on WYEN Request Radio these days, and it's all positive. That's because Jerry Mason, morning air personality and program director at WYEN, believes in positives. People don't need negatives, he says. They get enough of that every day. Radio should reflect the positive things in their lives; positive music and announcers who believe in good things happening. I think people identify more with that.

"Dance music was perfect for our format, a lot of positive wording," Mason submits. "The songs promoted a good feeling about radio and WYEN. An example was Al Stewart's 'A Song on the Radio' in which you feel the power it spreads."

On July 23, 1980, Walters wrote a reference letter for Mason and pointed out Jerry's success building WYEN's listening base.

Our ratings increased in women 25–49, from a .3 to a 1.3 in morning drive where Jerry was our morning person. Overall our station went up a full point in adults and women. We will miss his hands on attention to detail, programming work ethic at Y-107. I wish him well in his future endeavors and thank him for the work he has done.

Mason read the letter from Walters again after all these years. I asked for as much information from him as he could provide, and he came through with the reference letter, leading him to reflect that WYEN was a wonderful opportunity for its talent, a great proving ground.

"I will always hold my memories there in the highest regard."

Mason worked in the business of communications for forty-three years, almost half in Minneapolis, Minnesota, as an on-air personality for Viacom's WLTE-FM. Throughout his long career, he has been part owner, operations manager, program director, air personality, advertising executive, voice talent, and an audio and video producer. Today, Mason is managing partner of America Media Ink in Sibley.

Chapter Sixteen

STEP IN TUNE WITH THE WOMEN OF WYEN

In the 1970s, female voices aired with more volume than ever before. Yet, with a few exceptions, women in radio weren't waking up America in morning drive, the shift when people are listening in their cars heading to work. Women in radio tended to work in the evening, overnight, and on the weekend. Still, the women's movement of the 1970s brought women into a traditionally male-dominated profession. A number of professions like broadcasting saw a tidal wave of women while other professions saw a trickle.

Some of the best female voices treaded water in WYEN's talent pool. Val Stouffer, Jayne Neches, and Mary Louise "Louie Parrott" Parker worked mainly in the 1970s at WYEN, but Parrott stretched her WYEN career into the 1980s. None of them hosted a morning show, but women were making inroads in broadcasting, even as the first all-women radio station WSDM in Chicago grabbed headlines with such talents as Yvonne Daniels, Connie Szerszen, and Kitty Loewy. For a year and a half, Loewy hosted the WSDM Morning Show. She graduated from Southern Illinois University and had trained for radio at the Carbondale radio station WCIL-AM. Kitty admitted she leaped into Chicago radio. Her experience on WSDM in 1975–76, right after college, was atypical. Loewy acknowledged the all-female station was supposed to attract a male audience.

"Women were being hired up and down the dial in an enlightenment of the 1970s," Kitty noted in an interview for *The WYEN Experience*, yet she recognized it was still the exception for a woman working in a prime day part on radio. She couldn't answer with conviction why women were not working in the morning.

"Maybe it's the voice," she speculated, "or maybe it's that some of the guys are comedians." Loewy explained, "Typically more men in radio than women had a comedic bent, and perhaps this is what program directors looked for in their morning announcers. Program directors thought men were more appropriate in drive time than women because of their voice."

In the 1970s, Don Oberbillig was station manager of the most successful (based on revenue) "small market" station in the United States. Oberbillig recalled WSDR-AM in Sterling, Illinois, had received far more resumes from women for sales positions than for on-air jobs. He said the station wasn't receiving female applicants for announcer positions because there were not many women at that time trained for deejay work, although he did have two women news broadcasters.

Saying female voices weren't right for a morning or afternoon radio show wasn't a particularly sexist remark in the 1970s. Today, a program director would look for the best talent possible while the station manager might add "and affordable, as in best talent possible and affordable." Walters had specific ideals on female broadcasters. Jayne learned Ed was a traditionalist.

"He was very old school."

In a conversation with Ed, Jayne recalled her attempt at landing a prime shift.

She remembered him maintain, "Women didn't sound good during the day."

Neches didn't begrudge Ed for even an instant. She used the challenge by proving herself worthy of air-time. Jayne recognized that while Walters might not use women in the morning, he was willing to hire women broadcasters he believed were talented. Netches recognized Walters was ahead of the curve in hiring women, giving them a chance to improve their skills. Over a period of time, Ed modified his approach as he personally witnessed a vast improvement in the skills of women, which led to a prime on-air position for announcer Louie Parrott.

WYEN also benefited from Sherri Berger in sales. Shortly after Sherri started in 1977, she noticed all the spots (radio advertising) were voiced by men.

She pleaded with Mr. Walters, "This station is too male sounding. I think

THE WYEN EXPERIENCE

you need some female presence. Not every spot should be done by a man … that's a turnoff!"

Berger felt good that Walters listened and asked what she proposed.

"I said I did voice-overs in Kansas City and Wichita for years, and then I asked him to let me cut some spots."

He got her demo tape and immediately decided to let Sherri work part-time in voice production.

Today, you'll find women broadcasters on nearly every station across the dial in every radio market, working on-air and voicing spots. In a way, Jayne, Val, and Louie were pioneers, paving the way for a much more even playing field in broadcasting between the sexes.

Jayne Neches

Imagine talking one day to Mary J. Blige, the next day to Sting, and the following day to Janet Jackson. Neches's career track took her from working for WBBM-FM in the early 1970s to becoming a highly successful record company executive. In between, Jayne spun records on turntables for WYEN Request Radio. *Step in Tune* with Jayne over a nearly forty-year career.

For the early part of Neches's career, she and Jack Stockton worked at several of the same radio stations. Before Jayne began announcing at WYEN, she worked at WBBM-FM, where Stockton spent twelve years in CBS management at WBBM AM/FM in Chicago. Jayne worked as a secretary, but she clearly didn't stop there long. Her next stop on her career track was suburban WEXI-FM as news director, and she got there before Jack went over to the Arlington Heights station. But then Jayne moved to California briefly. Nearly a year later, she returned to the Chicago area and contacted Stockton, and he mentioned she should call Walters, which she did. Jayne landed a job as public affairs director. She covered town hall meetings. From her position handling public affairs, Walters gave her an announcer's shift, two nights a week and then full time, 10 p.m. to 2 a.m.

"Meier was on-the-air, 6 p.m. to 10 p.m. before me, so I became friends with Garry. He was one of the funniest people I ever met. He'd put up jokes in the studio, sometimes making fun of instances in the radio station or in the world."

A WYEN listener just requested "A Horse with No Name" on Jayne's show. Listeners would hear her say "Request Radio," and the calls would not stop. She recalled one very loyal listener who kept asking for "In-A-Gadda-Da-Vida," but Iron Butterfly's 1968 hit song wasn't one of the designated songs.

"Reynolds marked specific songs for play on the albums, such as Side A, cuts one, two, and three. I remembered one night I miscued a Ray Stevens album and put on his novelty song, 'Bridget the Midget.' People called all night asking for me to play the song again."

Jayne lived in Chicago with her parents during her WYEN years.

"I'm on-air one night, about to read some news, and I got a tremendous case of the giggles, so I turned the microphone level down, then thought I was okay, so I turned the microphone pot up and managed to get through the news. Although my parents couldn't always get WYEN's signal where they lived in Chicago, my father listened that night. He reprimanded me for not being professional. A few months later, we watched CBS's *60 Minutes* and famed broadcaster Lowell Thomas had died, and the segment about him showed bloopers about his work. He had a few where he just laughed and laughed. My father watched the segment and was he ever apologetic for telling me I wasn't professional."

While Neches worked at WYEN, she got invited to various record company events. She'd met local record promoters at WBBM-FM, and one time she remembered a particular promotion that stood out. She got to see the Captain and Tennille.

"Wasn't so much meeting them, but a representative for A&M Records told me about a new position and asked if I was interested."

Jayne went for the interview, and although she truly believed the WYEN job was one of her favorites, she thought she was never good enough. In July 1976, she went to work for A&M Records. Neches worked there for thirteen years, and before she moved on to Enigma Records, Jayne had been national sales director, head of singles for A&M Records. Naturally, national sales manager was next for Jayne. Unfortunately, she noted a juggernaut ahead of her for the position, so she went to work as a bigger fish in a smaller pond at Enigma as vice president of sales for a year, before taking a position as vice president of sales (de facto general manager) at a start up called Zoo Entertainment. But as this chapter is entitled *Step in Tune*, Jayne stayed true to her pursuit of the music business by becoming a senior vice president of sales at Geffen Records, and finally she stayed eight years at MCA Records as senior vice president of marketing and sales. In 2005, Jayne left MCA and started a record and consulting company, Artist Garage, with a friend. During those years with record companies from 1976 to 2005, Neches met the Carpenters, Nasareth, Janet Jackson, Styx, Mary J. Blige, Sting (solo), Police, was there at the recording of "We Are the World," and worked with Geffen Records at the time Curt Cobain died.

Neches is now retired, the final stepping stone in a path of great opportunities.

Val Stouffer

Through Facebook, I found Val. She crossed my path at WYEN for only a year, but some people need only a short time to make a lasting mark. Stouffer was one of those individuals with tremendous talent and a fabulous personality. After all, she *drank radio*. Val was impossible to forget. I let her tell her story.

The year was 1976. The United States celebrated two hundred years of independence from our cruel British overlords. The "Son of Sam" began a series of attacks that terrorized New York City for the next year. Steve Jobs and Steve Wozniak formed the Apple Computer Company. Viking I and Viking II set down safely on Mars.

I graduated from college and got my first job in radio at WYEN-FM in Des Plaines, which is almost Chicago.

I wasn't more than a month out of DePauw University when someone actually offered to pay me money to play music and talk on the radio. And it was a Chicago radio station. Well, almost a Chicago radio station. WYEN had a strong signal and was tantalizingly close to Chicago, which was the promised land for radio junkies like me who grew up in the Chicago suburbs.

I got hooked at an early age listening to the likes of Larry Lujack, Fred Winston, John Landecker, Bob Sirott, Dick Biondi, and Yvonne Daniels. Those guys were my cool friends, and Daniels got me thinking that maybe, just maybe, I could be on the radio too, if I could somehow manage to cultivate a voice and persona like hers. An FM station at the time, WSDM featured an all-female lineup of women with sexy, breathy voices.

I spent a good portion of my college years at the campus radio station trying to sound sexy on the air. I found it too difficult and humiliating to stay in character when I made a mistake on the air, which was a fairly regular occurrence early on in my career. Radio became a lot more fun after I quit trying to sound sexy and just sound like me. It turned out to be a smart business move too. More and more stations were adding a female voice to their lineups. Station owner Ed Walters called me and said WYEN's female jock had taken a job at a record company and he was looking for another woman to succeed Jayne Neches. I got the job working overnight for 120 dollars a week. I thought I'd died and gone to heaven!

WYEN's format was "Request Radio." We played listener requests as long as they were from the allowed songs in the music library. Our music director at the time was Rob Reynolds, who wasn't much older than I and was sort of a soft-rock, singer-songwriter visionary. He'd add an album and select various cuts from it that could be played during certain day parts. He was very particular about what songs were WYEN material and what weren't. Rob also did the afternoon drive shift. The rest of the air staff that summer consisted of Wayne Allen in morning drive, Garry Meier did middays, Rob, then John Zur, Chris Devine, and me. Dan

Diamond and his brother Pat had weekend shifts. A lot of us hung out together off-the-air too. We all pretty much lived, breathed, ate, and drank radio. We all dreamed of getting that big break at a "downtown" station. We were always sending out tapes and resumes. I was completely blown away when I got a call from WMET Program Director Gary Price in the summer of 1977 offering me a job. Garry went to the Loop around the same time. We both worked the overnight shift and often met for breakfast on Michigan Avenue after work. We remained friends for many years.

My year at WYEN was one of the most fun times I've had in the business.

Years later, when I finally got to meet the "First Lady of Chicago Radio," I realized Daniels was just being herself on the air too. The lady I met in person was the same lady I'd listened to and idolized and tried to copy such a long time ago.

Louie Parrott

The story of Mary Louise "Louie" Parrott can be told only one way ... as a team effort by her forever loving family. Just as I stepped aside for Val, I step aside for the family of the music lover.

I'm Louie's son, Robert Watson. Friends call me Dr. Bob. I feel the best way to do justice to my mother's memory is for my brother, stepbrother, and stepsister, and my father and stepfather to all work together and help me write my mother's story.

"You are the sunshine of my life," Louie sang as she headed out the door.

"Doo doo wah" was our response, even though we found out much later what the lyrics truly were. An infectious smile ensued, and off Louie went to entertain the masses at WYEN Radio. Louie was more than just an "anytime" disc jockey. She was also the charismatic Christmas hostess for the March of Dimes Telethon. She was the voice of countless radio and television spots heard and seen all over Chicagoland. She was a public relations liaison for the local Little League baseball organization. She was the quintessential hostess no matter whom the visitor. But most importantly, to the four of us kids, she was simply Mom.

Louise Parrott (Watson, then Parker) simply went by Louie. She had an incomparable zeal for life and everything that it offered, and she absolutely loved what she did at WYEN. Her one truest love was music, and for her to be able to share that with the world as well as be able to talk to the people she entertained, made for an enjoyable situation for her. No matter what was happening in her personal life, or if the weather was terrible, spinning the tunes, as she put it, was always an escape for her.

Louie's career started while attending Morehead State University in Kentucky. She was in marching band and played the French horn. In her music studies, she became interested in communications and quickly gravitated toward radio. She took a job at the university radio station WMKY, playing the hits of the late 1960s. She loved the fact that her friends could call her and ask to hear a song, and she could play it without question! To her, it was the job where she could entertain people and hear from them at the same time. Louie and radio was a perfect match. It's also where she met our father. After they got married, she went to work for the commercial radio station in Morehead where she had the morning radio show from 8 a.m. until noon. My father recounts an episode when the alarm clock went off at 7:55 a.m. one morning. Of course it was tuned to WMOR, and the early-morning deejay was chiming in with, "Well, I'm not sure where Bessie Lou is, but at 8 o'clock, this deejay is out!" Fortunately for her, living only five blocks away and not having a dress code to be a radio deejay meant that she was able to make it on time. But she absolutely loved what she was doing. She loved it so much that at one point she held a world record for consecutive hours of radio broadcasting at more than sixty hours! That record stood in the Guinness Book of World Records for almost twenty years.

Shortly after my birth in 1972, we moved to Chicago. My father had accepted a position with a very large advertising firm, and Mom had decided she would find a job in radio or television once she had arrived. Of course, back then there wasn't the crazy Internet to preface job listings and apply from afar. So when she arrived in Chicago, she stomped the pavement and cashed in a few friendly favors to land an interview with a radio station known as WWMM based out of Arlington Heights. She worked there for a little more than two years as an on-air talent. Through some networking and a few phone calls, she came in contact with someone who has been mentioned many times already, Ed Walters who owned a moderately well-known station, WYEN. He offered her a job to fill a time slot that had recently been vacated. At that time, I was still sporting single digits in age, but I still remember the day she pulled into the driveway and jumped out of the car to proclaim to anyone that was within earshot, "I got the job." The neighbors, who were out on their front porch partaking in a late afternoon libation, raised their glasses to her in a spirited salute. Thus began her career at WYEN.

Her main reason for the move was actually monetary, as hard as that is to believe. Louie had met many people that day, many of whom were friends of hers until her passing a few years ago. All of them spoke very highly of the job and the time slot as well as the ability to have some free reign in what she did with respect to playing tracks and running her show. She had a time slot that included part of the Chicago rush hour and an audience demographic that meant a significant increase in listeners as compared to WWMM. But it also meant that she'd be able to attend our baseball games, our piano performances, and other ridiculous kids' functions. Of course, being the new kid on the block also meant that she was the automatic fill-in for anyone calling in sick, needing vacation, or otherwise needing to miss their shift. I remember one time in particular where she worked her normal shift, made her way home to pick me up and take me to a baseball game on a particularly cold day. She then took me to my piano lesson immediately following the game, went home and cooked dinner for my dad and my brother, then came back and picked me up from my lesson. We got home, she sat down with me to eat dinner, and she headed right back out to cover a late-evening shift for someone who had called in sick. Frankly, I think she wouldn't have it any other way.

FROM THE WALTERS HEINLEIN COLLECTION OF PARROTT'S INTERVIEW WITH SINGER BOBBY VINTON.

One of the primary facets to the new job was a significant increase in recording radio spots. While she was well versed at recording her own spots, she did spend quite a bit more time recording before and after her regularly scheduled time slot. She recorded all kinds of commercial voice-overs. I suppose that spoke well for the sales department, but when her job description entailed a generous increase in time spent recording spots, it meant she would have to be a little more creative with her scheduling. That meant the kids got to go to the radio station a lot. Of course, for WYEN, that meant free talent. For us, it meant we got to hear ourselves on the radio and brag to our friends about it. I mean, how cool were we to have a mom

THE WYEN EXPERIENCE

that was on the radio, and that would bring us in to record radio ads as well? For kids quickly approaching their teenage years, it was quite an honor.

There was one specific incident that will forever be enshrined in my memory banks. As I mentioned, our father worked for a large advertising firm. This firm used to put on an annual breakfast, and he was charged with producing the event for many years. This particular year, they had managed to get a very quick shoot with the late Harry Caray, pregame at Wrigley Field. Coincidentally, Louie had an event there the same day doing a promotional special with the Cubs. Of course, what that meant for my brother and I was that we got to go to Wrigley Field and stand on the field in front of the Cubs dugout for the video shoot, meet Mr. Caray, then meet up with Mom and watch the Cubs play. This was a fairly exciting day for two preteens. Well, I believe the phrase "The best laid plans of mice and men often go awry" could not have been any more apropos.

The day started out normally. All of us (Mom, Dad, my brother, and I) left together rather early as there was a significant amount of setup time required for both events before the game. When we arrived, Mom went and met who she was supposed to meet, and we went off with our father for the video shoot. After what seemed to be several hours (to boys our age, anything that involves attempting to sit still in one spot for more than twenty minutes seemed like several hours), the time had come for Harry to come down and do his quick one-liner phrase, "I'm a 7-Up man." Harry expressed a lot of concern over the phrasing. His worry was that this video would strictly be used privately for this breakfast event and would never appear publicly. He certainly didn't want to upset one of his greatest sponsors, especially in altering their specific catch phrase. But after some mild coaxing and stern reassurance, Harry complied and agreed to begin. The bleachers and part of the scoreboard were in the scene as the point of view was eastward. Mom was sitting in the stands behind the dugout. She had completed her task for the day. Just as it is in the movies and behind the scenes, someone rolled out with the large black and white marker and said, "7-Up man, take one," then a short pause, and then, "action."

"I'm a Bud man!" Harry proclaimed in his usual manner.

"Cut. Mr. Caray. The line is, I'm a 7-Up man," said the set director.

Harry apologized and asked to do it again.

"7-Up man, take two, action."

"I'm a Bud man!" Harry proclaimed as if none of us were aware of the correct phrase.

"Cut, cut," the director said again, this time slightly agitated.

Harry dropped his head and said, "I'm sorry, guys. I know it's 7-Up man, but I cannot keep my eyes off of this beautiful woman sitting in the stands."

Just as if an eight-year-old pointed to the stands and shrieked, "Look," we all yanked our heads to see what was captivating Harry's attention. Harry was looking at Mom. She snuck down to those seats at some point while we were all engaging Harry in his plight for secrecy. However, Harry was the only one who knew this. The next few moments are a little blurry in my mind. I do remember someone saying, "I think he's looking at your wife" to our father.

Many kids may say they've been down by the side of the Cubs dugout for a game, but few can say they've been on the field with Harry Caray. I'd bet, though, we are probably the only kids who've been on the side of the dugout, on the field with the sports broadcasting legend, and had Harry hit on their mom! I don't think that memory will ever escape my thoughts.

I was ten years old when my parents separated. I reacted to the separation and eventual divorce by turning to my mom's WYEN friends. Though most of them were fifteen to twenty years older, my brother and I got to know them quite well. A few of them had the fortune of our entertainment on the nights when we were visiting Mom.

Annie Cothran of WYEN sales and her sister, Elaine, were my mother's intellectual voices of reason, but they also found time for me. They even taught me how to play backgammon, though it was a frustrating game for me. When the Cothran sisters came over, my brother and I knew this was a chance for Mom to vent, and she'd usually talk about money. She loved her job, but radio didn't pay well, and that was always an issue. Yet, we ate well, though I suspect Mom scrimped on herself during the week so she'd have enough money for us on the weekends.

Wally Gullick of WYEN news was just a great friend, but even more than this, he was a guy's guy, and I benefited from having him talk to me about things I wasn't comfortable discussing with Mom. He'd help me understand what I was thinking and feeling, a great translator with a fine sense of humor. I can still hear his voice today, nearly thirty years later.

Beth Kaye was in essence my mom's long-lost sister. Mom had two real sisters with whom she had a wonderful relationship, but they lived back near her hometown of Greenup, Kentucky. Beth was Mom's local sister. She'd always find the positive side to anything, and that was important as we went through this difficult time. I remember this one night that Mom's car was giving her a lot of trouble, so she was very late picking us up for our weekend, and she wasn't able to get to the bank in time to cash her check. I could see Mom on the verge of tears the entire drive from our house to her place about a half hour away. When we arrived, there was Beth with a rented VCR from 7-11, a couple of movies, and popcorn. She had spoken with Mom before leaving work and knew how important our time together was to her, and she just took care of it. I popped in a movie, and John and I sat down and watched while Mom and Beth went to talk in the other room. I could hear Mom crying and Beth saying, "At the end of the day, you've

THE WYEN EXPERIENCE

got your boys and a roof over your head." Beth always knew the right thing to say, and I know Mom was grateful for her friendship.

WYEN announcers Terry Flynn and Kevin Jay also played an important role in my mother's life, and there were others. However, I don't have specific memories on the others, and I apologize if I have left anyone out in particular. I know Mom was very grateful for everyone that helped her during those trying times.

It wasn't much more than a year later that Mom got remarried. I thought my new stepfather was a wonderful guy, and he had a couple of his own children. We weren't exactly the Brady Bunch, but with two Johns, a Kathy, and me, we were like the minor leagues for the Bradys. While the Parkers were not the fortunate beneficiaries of dugout Cubs tickets, they were quickly recruited as part of the voiceover crew. I still have a cassette recording somewhere that I made after we had all been down to record a spot for the famous Lattof Chevrolet. I don't even remember what was said or what our lines were, but I do recall having to do several takes because we were not accustomed to being quiet after we were done. Often our takes concluded with, "Well that stunk," or, "I sound like I had to sneeze." If we had just shut our mouths, we could easily have cut the amount of time in half reading the commercials. My brother, John, and I were still living with our dad and visiting Mom on weekends.

I suppose it's possible we took advantage of the fact we had an avenue of fame as a result of Mom's job and her employer's decidedly frugal finances. There was one incident in particular that certainly benefited us uniquely. My stepbrother and I were dating girls who were best friends. One chilly Saturday, we had scheduled an afternoon matinee bowling double date. Being Saturday, we were at Mom's house. Being under sixteen, we couldn't drive. Enter Mom who had graduated to a Chrysler Lebaron convertible soon after she remarried. For anyone who has ever ridden in one of these cars, they can relate to the fact Chrysler's claim that the car seated five was very loose. Sure it would seat five, perhaps four hamsters and a human. Well, we got to my girlfriend's house where they both were and realized we were not going to seat all five of us in this car. Mom suggested we put the top down! Never mind the temperature of close to freezing outside and we were going to drive on the highway for at least ten minutes. We anticipated frozen faces and numb limbs, but Mom turned up the radio just in time for us to hear my stepbrother talking to Mom in one of her radio spots. I think this temporarily took our mind off our freezing noses. And with Mom chiming in to mention that I also had a few radio spots, we were the coolest kids on the planet. Yes, we were pretty cold, but for that one moment, I think John and I had a sense of the joy Mom got every day doing her radio show. There was so much warm energy in the car.

As with all good things, Mom's chapter at WYEN ended, and she moved on to work for a small print production company in Chicago. A friend had told her to apply for this job and present some of her copy writing for some of her spots she had done. As it turned out, this was an excellent fit. Despite the fact the company looked at it as a junior-level job, the pay was still significantly higher than what she was earning from the radio station. That job got her a senior-level job with the March of Dimes, which ultimately led to her last job as art director for CTI Incorporated, a magazine company specializing in Christian magazines. Mom was proud of her faith and felt she had found her calling.

    I sensed a strong family atmosphere around WYEN that helped sculpt her life for years. She stayed in touch with many on the staff, perhaps not on a regular basis, but certainly in a capacity that she knew where they all were and what, if any, major changes had occurred in their life. Unfortunately, my mom lost her battle to cancer a few years back. As many of her former WYEN coworkers came to the funeral, we were reminded of the joy she had working in radio. Carol Walters gave me some pictures she'd been holding on to for nearly twenty-five years, which tells you what kind of family the WYEN family was in many ways. The poster boards plastered with pictures of her with all kinds of famous television, radio, movie, and sports personalities are a reminder of the wonderful opportunities afforded to her. My mother's infectious smile in each of the pictures is a reminder of how much fun it was no matter what the circumstances. Though she had left the radio world for other career experiences, radio was always a part of her. Her years at WYEN were some of the most cherished times that produced lifelong memories and eternal friends. We should be so lucky to have those friendships in our lifetime. Every now and then, I catch myself in a moment of deep thought reminiscing about childhood memories, and I still say out loud, "Doo doo wah."

CHAPTER SEVENTEEN

FARELLA AND BRYMAN

Donuts, Anyone?

He opened a box of jelly-filled donuts, fresh from his grandparents' Cassesse Bakery in North Chicago, and showed the delicious treats to the WYEN announcers gathered near the desk of front office receptionist Diane Finkler. Nick Farella was instantly likeable. The new WYEN jock in town came bearing jelly-filled donuts, and we did the expected when confronted with the freshest and puffiest sweet treats. Farella was a natural at tempting his coworkers. By the end of Nick's first week at WYEN, the staff outperformed Pavlov's dog, salivating at the mere thought of Farella's donuts on the front office coffee table.

I never forgot the first time I met Nick at the radio station in 1978. Donuts will do that to you—help you remember things. The impression Nick made on his very first day at WYEN wasn't just on me, but on the rest of the staff too. Some of us initially wondered why the new guy brought all those jelly donuts. I think we were attempting some restraint on our own by questioning Nick's motives. We learned rather quickly he only wanted to please his coworkers since he had access to fresh donuts every day from the bakery.

Nick's widow, Karen, knew Nick's life was so much more than sweet pastry. She called attention to his success at mentoring many broadcasters over the years.

Karen is a good-natured woman and offered depth about Nick's character that I surely did not know. We talked over the phone; she was in southern California where she remarried some years after Nick's death. I had to ask Karen how Nick chose radio when it seemed his background was in the family donut business.

"He would listen to the stations in Chicago and call regularly for one of the legendary jocks."

The jock was Dick Biondi on WLS-AM, and the two became friends. Biondi mentored Nick in radio and was successful in helping Nick realize his dream. So you might understand that on Nick's first date with Karen, he took her to visit Biondi, whom she quickly realized he couldn't stop talking about.

"I heard Dick on the radio, and he had this deep, booming voice, very commanding ... and we went up there and saw these young girls and guys all around the glass looking in, catching sight of Biondi. I couldn't see him, but Nick said Dick knew we were there, and just as Nick predicted, Biondi came out and started signing autographs. I was expecting this very tall, imposing figure, but instead, we saw this small, wiry Italian (laughs) that didn't fit his voice at all."

Nick's road to WYEN wound through Zion at WZBN and Waukegan at WKRS and WXLC. He didn't start as an announcer on WZBN; instead, he began working in sales and did quite well, so the bosses had him reading spots because they liked his voice. At the Waukegan stations, they gave him an afternoon shift. Then he went to Chicago radio, according to Karen, trying to keep track of all the moves Nick made before he came over to WYEN. Nick came armed. He had a first-class FCC operator's license and a box of donuts.

"Nick was really passionate about making the show the best every single time he was on-the-air ... and I know he learned this from Dick Biondi, though Dick was more wild and crazy. Nick was more to the letter and the purity of being on-air and giving information and sharing some time with his listeners to make their life a little bit better if he could. Nick was such a perfectionist of himself, and he did not fool around from the moment he entered the studio to the moment he turned off the microphone."

Karen set the perspective for Nick's close friendship with Wayne Bryman (Roger Leyden on WYEN.) The 1970s television show *Odd Couple* offered viewers an amusing interaction of mismatched roommates—actor Tony Randall as Felix Unger and actor Jack Klugman as Oscar Madison. Felix was fastidious and immaculate in every aspect of his life, while Oscar was the opposite. Not that Nick was the perfect match for the character of Felix, but Nick had his moments, especially in his relationship with Wayne, a pretty good Oscar who always found a way to get to Nick at WYEN. The two had a special working relationship Karen couldn't describe without smiling.

"Wayne was such a good guy, using comedy as a stress release, and he was good at it for the staff, but not with Nick."

On his first day at WYEN, Wayne noticed he could see all the way through the production room into the on-air studio just by standing in the reception area. There he spotted Nick preparing for his afternoon show.

"Nick had the most angry look on his face; he would not smile." Wayne laughed, reflecting on their *Odd Couple* connection.

"I said, 'Nick, you are doing afternoon drive. You are supposed to be bouncing off the walls; giddy and crazy.' Here he is looking so serious, and I made it a point to try and make him laugh and smile."

Wayne couldn't crack Nick's hard shell. Nick, though, reminded Wayne of his rule.

"There is no belching, farting, dumb jokes; nothing in this studio until 8 o'clock," Nick insisted.

Chuckling softly, Wayne smiled at Nick, shaking his head and opening his hands. "Nick, that's my whole opening act!"

Billed as Request Radio, WYEN encouraged listener calls to their favorite announcers. But this was more than merely requesting a song; they'd talk about their lives and interests. Not unlike a customer at a bar talking to the bartender or a patient talking to a psychologist, the announcers and their WYEN listeners built a special bond. Similar bonds were established at other stations as well. This suited Nick. He was definitely people oriented. Were the social networks created during Farella's day, he'd have used everything available for building his audience. Instead, he used what was available, mainly public events and working the phones.

Karen says Nick shared with her the fact he had a large female following.

"So-and-so called and wants to have drinks," Nick would tell Karen. "I told her I'm a married man, but she keeps calling anyway."

Karen didn't have trouble sharing a part of him that enjoyed connecting with people.

"Working the evening or overnight shift, Nick would often times talk to lonely people and bring a bit of sunshine into their life."

Nick told his wife about a female listener he talked to on the phone that he thought might not take the hint he was married with children. Her interest in Nick came clear one day just before Christmas. Nick brought home a huge box of decorated gifts for the entire family from this persistent female WYEN listener. Karen had no problem with this gesture of friendship. She felt the listener appreciated all the time Nick talked to her on the phone and enjoyed his show and wanted Nick and Karen to understand she cared enough to do something special for them.

No question Karen was Nick's number-one listener. She'd listen all the time.

"I'd be with my girlfriends, and I'd turn on the radio in the car, and they'd ask, 'Who is that?' I'd say, 'It's Nick,' and they were okay with that, and then they'd come over to the house, and I'd have the radio on in the kitchen for when Nick would speak."

Karen listened on that fateful day in May of 1979.

"A news flash came on briefly alerting us to a horrible plane crash out of O'Hare."

Karen ran to the phone and called Nick, and he said, "I know, I know, I heard." But then he asked who Karen had on, and he urged her to go and listen to this TV anchor and that TV reporter and asked her to call back, letting him know what they said so he could share this information while he coanchored with Wayne. "Not an understatement describing this as the most intense Nick and Wayne ever were together," Karen asserted.

Announcers have the satisfaction of pleasing their listeners with music. Songs that spark memories of good moments, songs that make you tap your feet or move your body, and songs you sing aloud are what good announcers hope they'll elicit from you as you listen to their show. Nick's audience did all these things, and the proof is still alive in all the tapes (airchecks) he made over the years. Although Nick died in 2005, Karen has dozens of tapes, but at first, she could not listen to them.

"Maybe seven or eight months later, I listened to an aircheck, and it was really hard, but it made me feel good, but also like my heart was breaking again, and I'm starting to get that way again talking to you."

Karen paused for a moment.

"It brings comfort, and it's good because it's a tangible connection to him and it's a beautiful thing."

A clearly sensitive moment for Karen telling me she had to wipe tears from her eyes. She wishes everyone could have tapes of a loved one because it brings comfort and joy, though it's probably different for everybody.

"To be able to hear the tape and say to your kids or have your grown children eventually say to their own children, 'That was your father (or grandfather),' and they look at you with wonder, and you show them a picture. That is a special moment, an affirmation of who he was."

Nick was in his midfifties, and he and his wife, Karen, and kids were having a great time in their lives. Karen and Nick never looked at the years, never cared to comment on their age.

"Nick lived life to its fullest, and he managed to make two things work—

his love of his family and his love of radio, and he managed to marry those two together."

Karen fondly remembers how Nick enjoyed his can of Coca Cola. Anyone who'd come by would be offered the soft drink, and he'd give them money, and they'd sit together drinking Coca Cola. He also got a kick out of mentoring young people and seeing them succeed in radio or television.

"He must have been a teacher in another life or something," Karen quipped, "even finding them jobs he'd say was good for them."

Asked for a list of some of the broadcasters Nick helped along the way, Karen said the list included a couple of people on local television today.

Long after Nick left WYEN (1978–1983), he'd been working as the afternoon personality at WERV-FM (95.9, The River). Wayne and Nick were talking about the great White Sox team (World Series year) as Nick was a huge Sox fan. Nick told Wayne he wasn't feeling well, so they went upstairs for some tea, thinking Nick had a sour stomach…but Nick never had a chance to drink the tea. Someone called 911 just as Wayne's son, a firefighter paramedic, pulled up and found the scene with Nick on the floor, and he took over.

That was on September 9, 2005, a day Bryman says he'll never forget.

"Nick literally passed away in my arms."

This shook Wayne. He instinctively looked for his desk, found it, and screamed at the top of his lungs.

"I freaked out … you know, we bury our grandparents and parents, but we're not supposed to bury our friends yet."

At the funeral at Strang Funeral Home in Grayslake, Karen learned something about Nick that told volumes about what was in his heart. Nick had never mentioned to Karen his little ritual nearly every morning on his way to work.

"A young man," Karen says, "came up to me at the funeral. I knew he was mentally challenged, and he told of how he'd walk to the grocery store where he worked, but almost every day Nick would drive by, stop and pick him up, and drive him to work. But that little revelation only warmed my heart more because the young man held my arm and looked directly at me. I could tell something that would surprise me was forthcoming."

He said, "Nick would come back to the store and take me home."

The suddenness of Nick's death shocked his friends and brought many of them together for his funeral—an image Karen felt showed tremendous respect in the life he led.

"They told the same old war stories," but this offered a kind of peace because this was his world.

Sometimes Karen feels she needs to sing a certain song that makes her feel calm. The song is "Above the Storm," first sung at Nick's wake by Jimmy Peterik of Survivor.

"I still know all the words and sing it when things get a little scary. I guess that is what radio does with the music and talk. It brings us a little relief in our everyday lives."

Farella and Peterik were good friends.

"He sang this beautiful song he wrote with his friend Johnny Van Zant of Lynyrd Skynyrd years before, but never finished," Karen vividly recalled. "Peterik finished the song with some help from Nick up above the night before the wake."

The songwriter for "Vehicle" by the Ides of March and cowriter of "Eye of the Tiger," and a song Nick loved, "LA Goodbye," now had the inspiration to complete "Above the Storm," though he would have wished for it to have been some other way.

FROM THE WALTERS HEINLEIN COLLECTION.

"I remember as a child, just sitting on the porch ... as the rain poured in and the thunder rolled ..." Peterik sang, and all those who'd gathered at the wake sat still. Their tears dabbed slowly, shoulders rounded, heads bowed,

and the music spilled out. *"Telling us to hold on tight, now when the storms of our misfortune are more than I can bear."* Some looked toward Karen and her family, hoping for her to bear up under the strain. *"The sun's gonna shine, the rain will stop, the healing winds gonna feel your heart ... my love will keep you safe and warm."* Those straining their eyes for one moment from the back or in the middle of the room could see and hear Karen singing too, for she instinctively knew what Nick wanted. *"...I carried you above the storm."*

Farella's radio travels stopped at WYEN-Request Radio, WKQX-FM, WXLC-FM, and WERV-FM. Nick passed away in September 2005.

Roger Dodger

In the world of Wayne Bryman, it's not a cliché to say "he got to live the dream."

In the 1960s, a young Wayne looked through the WLS and WCFL observation windows at legendary announcers whose voices vibrated out of our transistor radios. Radio was *it* for Wayne. He connected with the music, wit, and humor on the other side of the glass. Yet looking at the announcers and hearing them perform weren't enough. Wayne's grandparents bought him a tape recorder, stoking the fire within. The boy was already listening to radio stations and absorbing the music and jingles; now he played deejay in his room with a recorder.

Forward twenty-one years. A song ends, and the announcer switches on the microphone, and the WCFL live studio light goes on. Wayne's voice, deep and resonant, booms from the speakers in the observation room, and the kids looking through the glass are smiling back at him, a few even wondering whether they'll be the next *Wayne Bryman*. Wayne knows the look, for he had a similar one wondering whether he'd be the next Dick Biondi.

"It's like a kid who wants to be a Major League baseball player, and then all of a sudden he's on the field, wearing a uniform, hat, and holding a bat or glove."

Bryman worked at several major Chicago radio stations, but his feeling about WCFL in downtown Marina City was special. He could finally answer his own questions. What was behind the observation window and was the whole dream worth it?

"My first day on the air at WCFL, after playing a record or two, I looked over my shoulder and realized I was on the other side of the glass, not a little kid watching Joel Sebastian or Ron Britain."

Tempered with a child's sense of awe, Wayne remembered some words of advice he received from legendary announcer Fred Winston.

"Fred told me it doesn't matter where you work; the BS is all the same. The only difference is the size of your paycheck."

Winston cut through pretty quickly, it seemed, thought Bryman.

"He was a hoot to work with at WCFL. I also worked with Dean Richards, and he was one of the great guys who always landed on his feet."

Friendships at WYEN didn't mean any less to Wayne than at the Chicago stations.

"I enjoyed WYEN; though we didn't make any money, we did make good friends."

He became great friends with Nick Farella and Mike Roberts, mostly because he saw them so often following Nick's shift on-air and finishing his shift before Mike went on.

"Nick was one of those people despising country music, and during the Urban Cowboy craze of 1979–1980, songs crossed over. Smash hits on country crossed over to pop like "Looking for Love" by Johnny Lee and "Stand by Me" by Mickey Gilley. But Nick refused to play these hits. My first hour after following Nick, I'd have every country cross-over hit. Years later, Nick loved the TV show *Reba*, so you figure."

Wayne left WYEN for Satellite Music Network. He equated his five years at SMN in Mokena, Illinois, to one big country club. He lived in Hoffman Estates and drove on Route 83 all the way to Mokena, where he'd find Brittan and Mike Tanner of SMN in the building. That's where he announced overnight Country Coast to Coast, which his sister picked up in Las Vegas.

He'd record his kids, and they'd say cute things like *my mommy lets me stay up late listening to my daddy Wayne Bryman*, and his sister would hear that stuff, and it would make her homesick.

"I had a great ride though I didn't become another Biondi, but that doesn't happen to most of us."

Chapter Eighteen
AMERICAN AIRLINES CRASH

Wayne Bryman (on-air as Roger Leyden) never anchored an internationally important breaking news story. He was a WYEN announcer, subbing for a vacationing news anchor on May 25, 1979.

Robin Pendergrast never rushed to the scene of an accident where he felt he could not save a life. On that date in May, the volunteer firefighter paramedic worked his main job as a public relations executive.

At the moment Wayne and Robin went to work, each man had no idea how challenging a day they'd soon face.

In the middle of the afternoon on what was shaping up as a sunny and warm day, a fully loaded DC-10 aircraft crashed into a field near a runway at O'Hare International Airport. Two hundred and seventy-one passengers and crew on board and two people on the ground died in the explosion of American Airlines Flight 191. The Federal Aviation Administration investigated the crash until they had the cause well documented.

Many people driving near O'Hare saw the plane in the air, and still more people saw a huge ball of fire shoot up into the air.

Emergency personnel rushed to the scene of the crash in an open field, near a trailer park, not that far away from the WYEN studios.

Wayne did not see the plane in the air, nor did he see the crash and fireball. Robin saw the fire and smoke. This is the story of how they each handled the crisis separately.

A professional photographer, Pendergrast, is always ready with a loaded camera. He'll take pictures of events unfolding, weddings, parties, and in-studio candid shots. He took to photography at the age of eight, and nothing since had really challenged him, not until that day in May.

Bryman can talk about anything, and you need not wind him up. He's ready to go and go, and he doesn't tire, nor do the people listening to him. But can a jock switch to news on one of the most challenging days in the history of broadcasting?

Today, Robin owns his own studio, RFP Photography, but back in 1979, he was a public relations executive and a volunteer with the Northfield Fire and Rescue Squad. The only camera in his car was loaded with slide film on May 25.

Wayne had a regular late-day DJ shift on WYEN, but on May 25, the entire WYEN staff would stop what they were doing and become a news-gathering team, feeding information to Wayne.

Robin and Wayne went about their business while hundreds of passengers boarded a DC-10 airplane at O'Hare Airport bound for Los Angeles.

Allstate Insurance Company in Northbrook generously donated thousands of dollars for outfitting motorcycles with medical supplies. Working public relations, Robin's client was Allstate on that day in May. He was supposed to help Allstate kick off Northfield Motor Medics.

WYEN needed a fill-in for a vacationing news anchor, and Wayne was the guy. He normally worked the 8 p.m. to 1 a.m. jock shift, but Bryman, the 8 p.m. to 1 a.m. announcer on WYEN, filled in as the station's news anchor for a vacationing news person. He was ready to read the 3 p.m. news on the Nick Farella Show.

The DC-10, fully loaded with passengers and fuel, taxied on a runway at O'Hare.

United Press International (UPI audio) started after the 3 p.m. station identification, and Bryman (Leyden) sat down at the WYEN audio board. Farella left the room. He picked out records for the 3 o'clock hour of his show.

Pendergrast had everything in place for Allstate's announcement of how it would outfit motorcycles that paramedics could ride to accidents on the Edens Expressway. Road crews working the Edens tore up fifteen miles of the expressway, leaving four lanes of the Edens as dirt before they began reconstructing the road. Only two lanes were left for traffic, making it near impossible for an emergency vehicle to quickly reach the scene of a crash or other type of emergency.

"You'd have to climb over fences to reach anyone," emphasized Pendergrast.

At 3:02 p.m. on the Friday before the long Memorial Day weekend, Flight

191's pilot, Walter Lux switched on the engines, and they roared as he guided the DC-10 down the runway. Two hundred and fifty-eight passengers relaxed in their seats, watching on a closed-circuit TV screen the plane taxi down the runway as the pilot and copilot saw it from the cockpit. Thirteen crew members were all in position for a four-hour nonstop flight to Los Angeles.

With his copy in hand, Bryman was ready to turn on his microphone.

Pendergrast had a stake in the press conference because he knew as a captain of the Northfield Rescue Squad, he qualified as one of the many emergency responders trained on a Motor Medic. Chicago television crews, Motor Medics, Allstate executives, and others crowded into the Allstate press conference.

The pilot had the plane in the air, but the FAA said the DC-10s left engine loosened from its pylon assembly and shot forward and backward, severing the hydraulic lines, causing a hydraulic leak onto the runway. Despite his 22,000 flight hours as the pilot of DC-10s, all of Lux' experience would not help him see what had happened to the airplane. However, the tower control saw parts of the pylon assembly fall away, landing on the runway. The airliner drew closer to takeoff, and then the engine shot forward and backward and out of its support. The investigation described the horrible development, but the tower too may not have recognized the extent of the damage, saying, "American 191 heavy, you wanna come back and to what runway?"

UPI News ended, and a spot (commercial) played. Then Bryman cut out the rest of the network newscast as was expected and read local and regional news. He noted the time as 3:03 p.m. because he'd have to finish at 3:06 p.m. and return the station to the Nick Farella Show.

From the pilot's vantage point, it was impossible for Lux to realize the number-one engine actually fell off the plane, even as he looked back toward the wing. He would not have had a visual advantage from the cockpit. Not only could he not see that area of the airplane where the engine was missing and the wing slats were slowly retracting, but his control panel became disabled. Lux acted as though the engine just stopped working, and the plane continued climbing. He followed the DC-10 book without one of its engines working. No one in the cockpit responded to the control tower asking whether AA 191 wanted to turn the plane around. Lux and his copilot were just too busy flying the crippled plane.

Fifteen seconds into the Allstate press conference, Pendergrast heard someone say, "Look to the left!" Through a large window, Robin saw a big black plume of smoke.

Maybe it's a house on fire, he thought, but suddenly Pendergrast heard a series of very familiar sounds.

"All the pagers went off," he remembered, and moments later, "the entire press corps left the room."

Pendergrast stood stunned. So were the executives of Allstate and Motorcycle Medics. "Someone came up to us and said an American Airlines DC-10 crashed off of Touhy Avenue in Elk Grove Village."

Talking about this so many years later, Bryman felt for a moment he was back in the WYEN on-air studio running the console, taking a call from a WYEN salesman who saw the plane crash from Arlington Park. Program Director Jerry Mason worked to help organize the team of sales people and office staff. They brought information to Wayne and Nick, and the two announcers stayed on-the-air for five hours of non-stop coverage.

"Ed or Carol told us we beat WBBM-AM, but we don't know, just rumblings to that nature."

Wayne could never be sure, but whether first or last, he knew he gave everything he had into the reporting.

Instinct kicked in; Pendergrast got into a backup ambulance and headed toward Elk Grove Village.

"Back then, there wasn't the traffic control of EMS, Emergency Medical Services, there is now. We saw hundreds of emergency equipment, and it was chaos! That's one of the reasons for the evolution of MABAS, Mutual Aid Box Alarm System."

Pendergrast described the scene he encountered.

"People were running down Touhy Avenue toward the crash scene. However, police had already cordoned off the site, letting ambulances and fire trucks through."

Pendergrast got there in fifteen minutes. The sixteenth minute was life changing.

"The plane was in pieces, bodies everywhere. I could smell burning flesh. The air was black with smoke. Steam rose from the fire."

Around the burning fuel, black smoke, and steam, firefighters laid out four-inch supply lines. Every moment the fire raged, they worried about a nearby gas storage facility.

He knew the ambulance teams were not finding any survivors and realized he probably had no business being there; many didn't, he reasoned. Pendergrast grabbed a camera, checked it for film, and then realized he had not packed high-speed film but less than ideal slides. Didn't matter. He began shooting the crash site anyway.

"I took some shots of victims burned alive, but I never sold the photos or took them anywhere. I knew the photo editor of *Time Magazine*. Suzy Kellet was negotiating with a guy who got the shot of Flight 191 through the window at O'Hare. The shot showed the plane's left wing dipping down, the plane nearly perpendicular to the ground. He was negotiating buyouts of that photo and a shot of the huge fireball for five thousand dollars. I had some discussion with Kellet later. I told Suzy I wouldn't sell any of these crash-scene photos. That would be wrong, partly because I was on duty as a paramedic, and capitalizing on something like that is inappropriate."

What Pendergrast recognized was the crash scene had two different control points. Many of the ambulances were arriving under the control allegedly of the Elk Grove Village Rural Fire Protection District. But on the other side of the crash, the City of Chicago had their own patrol points set up. He believed there was no collaboration between the two.

"I did this story for *Fire Chief* magazine, and if something positive came out of this tragedy, it was the City of Chicago realizing they needed to coordinate with all the suburbs, whether it was Park Ridge or Elk Grove Village or anywhere else. Very elaborate systems are now in place. Back then, no one ever assumed something like this would ever happen."

Wayne and Nick finished their coverage at 8 p.m. Wayne knew he did everything he could to bring the story to the WYEN listeners. They heard eyewitness accounts, repeated what they understood of the crash, told of what family members could do for updates, explained the type of aircraft in the crash, and offered people an opportunity to talk.

Chapter Nineteen

A SMILE IN THEIR VOICES

Satellite Tanner

Bryman hugged his family at the end of the day on May 25, 1979. He finally came home from WYEN after hours of broadcasting information on the American Airlines plane crash near O'Hare. The crash had shaken him with such loss of life.

I came home to my wife and young son and hugged them both the evening of October 25, 1995. Seven students at Cary Grove High School died in Fox River Grove at a rail crossing we came to know as The Seven Angels Crossing. I had reported all day from the site of the tragic bus/train crash.

Mike "Satellite" Tanner and his family and millions of others across the country cried on September 11, 2001. Mike hugged his loved ones and felt the spirit of a national resolve that this nation would not fear terrorism but dedicate itself to collectively fighting against it.

Wayne, Mike, and I shared a bond as broadcasters. We maintained our composure in the chaos. But privately, we felt what you felt, and on those dates of extreme grief, we were certainly put to the test. Tanner was exceptionally talented in both areas of on-air, announcer and news.

Mike has this incredibly pleasant voice and makes everything sound like things are good with the world. This ability made him perfect for announcing, in which he was clearly a bright star. On LinkedIn, Tanner is listed as having worked as a Network Adult Contemporary Air Personality/Program Director for Dial-Global Communications (Formerly Westwood One Radio/CBS Radio/Unistar/Transtar Radio) from August 1985 to October 2008.

I didn't know he had a television career as a child, but through some research, I found that he appeared on *Dragnet* and *Marcus Welby, M.D.* and came close to television immortality. He was a finalist for one of the children on the *Brady Bunch*. On LinkedIn, he'd been listed from 1959 to 1976 as a child actor and voiceover actor at various agencies and studios in Chicago and Los Angeles.

After WYEN, I lost track of Mike. He worked there from December 1977 to November 1978. We were both so busy with our careers, trying to spend as many hours as possible perfecting our craft. Funny how just one thing can bridge the distance in time, and that one thing happened as I drove back to Crystal Lake, Illinois, from Springfield where I had attended the Illinois Broadcast Association Spring Convention in 1987 to accept the Silver Dome Award for a public affairs program. I chose the lesser traveled roads driving home the three times I did that in the mid-1980s. Yes, the trip was longer, but I got to see central Illinois for all its feed silos and cornfields. What I used to do was tune into radio stations so I could hear if the station was doing something local that stood out. I happened to tune into a station in LaSalle Peru, Illinois, as I drove north on Route 51 and heard someone with a familiar voice, but I couldn't place it. This was a pleasant and clear voice, knowledgeable, and he seemed to talk directly to me. I waited for his name and eventually heard the name Mike Tanner. Wow, I had worked with an announcer on WYEN-FM ten years before, and he did sound similar. But I couldn't imagine he'd been working in central Illinois because I figured with Mike's talent, he'd have landed a job in New York or Los Angeles or even Chicago. But it turned out Tanner's program was heard around the country. Mike was one of the most successful announcers I'd ever worked with in radio and one of the nicest guys around. I asked Mike to fill in the spaces of his career by writing his story.

During my thirty-two years in broadcasting, I've recognized the New York radio market as number one, but in my humble opinion, the Chicago radio market should be number one as far as the home of great talent and creativity. After a short-lived television career in Chicago and Los Angeles, I was fortunate to start learning about the medium of radio while I attended New Trier High School. This was where I met up with Rob Reynolds, who worked with me at WNTH, our high school radio station.

After graduating high school in 1973, everyone went off in different directions and paths. While attending four years at Loyola University in Chicago, I got more involved with all aspects of radio. I worked on-the-air for the college radio station and had the opportunity to intern and work part-time at WLS Musicradio doing music research and programming with John Gehron, and then a couple of years later, I worked with John Platt and Harvey Wells at WXRT Radio. These incredible experiences started shaping the direction and career path that I wanted to follow. (Starving, but having fun in radio.)

As I was closing in on graduating in 1977, I turned on the radio and heard Rob, my old high school friend, doing middays on WYEN Request Radio. I called him, and he suggested that I send a tape to Ed Walters. After receiving my first rejection letter, and since Gehron didn't have a full-time gig at WLS for me when I graduated, I thought law school would have been a better idea.

Gehron put me in touch with a WLS sales manager who built a station in Rockford. He recommended that working there doing a little bit of everything would give me a lot of experience, and after some time on-the-air, maybe I could even come back to WLS when Bob Sirott left one day. (Wishful thinking.)

After six months at WYBR, I applied to stations all over Chicago. The program director at WYEN had an immediate opening for a morning newscaster. I gave notice in Rockford and took the job as Scoop Tanner at WYEN.

I confided in WYEN Program Director Wayne Allen that I hoped to do an air shift again. He came through after a short time, putting me on the all-night shift, and I gave up the news and traffic gig. It takes a certain type of person to stay up all night, and I knew that wasn't me. Fortunately, the afternoon drive shift was opening, and Wayne moved me to my new home.

On a clear day, WYEN's 50,000-watt signal reached four states! This was a dream come true. I followed Rob on the on-air schedule just like our high school days. We were all part of the Request Radio family with Ed and Carol Walters at the helm. We followed a color-coded music clock, logged our commercials, took transmitter readings, answered the request lines, produced commercials, cleaned the heads on the cart machines, cued up records on the turntables, and spliced tape, all without the help of computers or producers. With the advancement of computers and audio editing programs as well as MP3 files for recording voice and music, the era of real hands on radio disappeared. Also gone is the opportunity to be able to work at a major market station and learn all aspects of radio.

WYEN always had a professional team on-the-air and behind the scenes as well, including a news staff headed by Stew Cohen. Stew was the news director and afternoon anchor on my show. It was a Friday night, May 5, 1978 when we decided to get together after the show. A local disco, the Savoy in Wheeling, was the place that the WYEN jocks would take turns during the week spinning records there at night. Rob was in charge of the disco dance music that night. Stew and I asked a couple of ladies to dance. While on the dance floor, Rob proceeded to introduce us as fellow coworkers at WYEN. The person I was dancing with

THE WYEN EXPERIENCE

ended up calling me on the request line the next afternoon, and we went out that evening, and I brought her with me to emcee a wet T-shirt contest. The contest was one of those hosting duties that were part of my job at the station. Anyway, Marla married me in spite of those hosting duties, and we've celebrated more than thirty years married.

"Anyone for football?" WYEN announcer Mike Tanner snaps an album to Bob Walker.

Shortly after leaving WYEN, the disco years faded, and the 1980s began a different era of radio. AM Radio was the new home for talk and news formats, and FM Radio was home to a variety of different music formats with satellite radio looming on the horizon.

Fastest on Track

"I'm dangerous with a live mic!"

Between calling races one day, the *Voice of Maywood Park*, Tony Salvaro, didn't know his microphone was still on.

"I yelled out to the press box ... asking if anyone is going down to make a bet. The crowd roared. Someone came into my office and told me my mic was on. All I wanted to do was place a bet!"

Salvaro was not only the voice of harness racing at Maywood Park, he also produced the Maywood Park Racing Wrap-Up on WYEN.

Compared to the announcers and newscasters, Tony was every bit just as significant in his role at WYEN, despite his rare visits to the radio station. He phoned in all his Racing Wrap-Up reports.

Salvaro was a colorful character in the sense he fit in so well with all the famous people visiting Maywood Park during his career there. *The WYEN Experience* would not be complete without a visit to the voice of Maywood Park and his remarkable career.

At the age of twenty-seven, Salvaro was the youngest in the nation calling races, and this excited the owners of Aurora Downs, where Tony was completing his first week of work. The track owners anticipated great things from their race caller. Judy, the switchboard operator at Aurora Downs, called Tony and told him the owners were listening to his calls and could not wait for the big race, which they all knew was next.

"Now I have to put pressure on myself and really do a good job. I asked myself, What would Phil Georgeff do to make this race special?"

Georgeff was a nationally famous race track caller at Arlington Park and gave some career guidance to Tony along the way. The big race was called The Geneva, for Geneva, Illinois, so Salvaro had this great idea similar to something Georgeff might say.

"And they're off in the Geneva."

Salvaro was pretty proud of himself with his catch phrase. The race started. Tony leaned in and exploded into the microphone.

"And they're off in the vagina!"

Salvaro wasn't sure what he just said, but the crowd reacted immediately, he recalled.

Tony heard them roaring non-stop, and all he could think was, *Tony, you're fired*.

He thought for a moment that maybe no one heard him in the owners' box. He knew they were having a great time ... then the phone rang!

Oh, my God, Salvaro thought.

Tony held his breath, thought the worst, but he suddenly recognized the voice, Judy, the switchboard operator, so he relaxed a bit, feeling some level of relief.

"Tony, you made me pee in my pants," she confessed.

"I just crapped in mine," Tony admitted.

Judy sent Andy Frain ushers down to the table where the owners sat. They found the area crowded and loud. You just couldn't hear anything, but the owners were having such a great time anyway. Salvaro never forgot that he barely escaped fallout from his first big mistake in his rookie week.

THE WYEN EXPERIENCE

In my interview with Tony for the book, I asked him about "Here they come, spinning out of the turn." This was a wonderful catchphrase that floated right off the tongue, but it was not Tony Salvaro's; it was Phil Georgeff at Arlington Park calling the races. I knew this catchphrase well because the phrase was always the highlight of racing results at Arlington Park. I figured just about everyone in Chicagoland remembered Georgeff's famous line. Tony knew his success in race calling depended on developing a unique style.

"Here they come turning for home."

Before I went home after finishing my evening news shift, Tony would call from Maywood Park and read the Racing Wrap-Up over the phone. I'd tape his report and pass it along to the announcer for playback at prescheduled times late in the evening. Racing Wrap-Up was among the most popular programs we ran on WYEN. Thousands of people tuned into WYEN for racing results at Maywood. This was incredible, but a testament to Salvaro understanding what people needed and the wisdom of Ed Walters letting Tony report the results on radio.

Patricia Salvaro, Tony's sister, actually deserves credit for matching Tony's expertise as a race caller to Walters's WYEN. Patricia lived upstairs from Tony, as he tells the story.

"I had an idea for racing results on radio. At the time, only thoroughbred results were available, but no one would touch it on television or radio. I was mentioning this to my sister, and she said, 'Why don't you try the radio station I listen to all the time, WYEN?'"

He wasn't one to procrastinate, so he left the room and called WYEN, and Walters answered the phone. Tony introduced himself as a track announcer at Aurora Downs and said that he had an idea to put racing results on the radio. Tony remembers this was March of 1975, and Walters was quite receptive.

"Could you tell me what it's going to sound like?" Walters asked Tony.

Salvaro was willing to give him a fictitious report, basically one or two minutes, and show him not only his voice and information, but the cadence of the delivery.

"As soon as I'm done, Walters says, 'That's great,' and asks how soon I can start. I started four days later with six reports a week, featuring my race calls and interviews with harness drivers and celebrities at the park."

From Aurora Downs, Tony moved his race calls to Maywood Park, and that's where he probably had the moment of his career. Maywood Park was honoring Muhammad Ali, and he was there for a charity event.

"Ali came up to the booth to meet the announcer, and sure enough I talked him into calling a horse race."

The Ali event was so big that Wally Phillips of WGN 720 called the next morning and asked for the tape of Ali calling a race. Walters heard Racing

Wrap-Up and an interview with the boxing legend and was thrilled with Tony's celebrity interview.

"Ali had the quickest mind of anyone I'd seen. He stood right next to me with a microphone, and I told him he'd have to say whatever I said in his ear, and he said, 'Go ahead,' and the horses were off." Tony was giving the names of the horses.

"My God, Ali's mind was so quick; no sooner did I get the words out of my mouth, but they came out of his mouth. This guy was magnificent, the best, and you could hear the crowd appreciate the champ calling the race."

Ali stayed in the booth for a while because he liked the view of the track and being able to talk to the crowd.

"He wanted to give out autographs and borrowed a pen from me and began signing papers the size of a waitress's writing tablet."

Ali looked through the window, saw thousands of people below, four floors down, looking up at the booth.

"Ali says on mic he'll throw his autograph down to them, and the papers float in the air, and then he says he's going to throw out one-hundred dollar bills, and the crowd roars."

After leaving WYEN in the mid 1980s, Salvaro went to WBBM All News Radio. Salvaro says Chris Barry, then news director, hired him after Bob Feder said in his column it was a good idea. Salvaro stayed for eleven years with WBBM.

Tony's dream of calling races began in high school, became reality in the early 1970s at Aurora Downs, and ended in 2002 at Maywood Park after twenty-three years of calling races there. His love of horse racing and calling the races never wavered, but his health became an issue, forcing his retirement.

Chapter Twenty

A SIDE TRIP TO PLAINS

Shortly before the presidential inauguration of Jimmy Carter, his brother stepped into the role he was born to play. Billy Carter started entertaining the country with his Southern charm. Reporters in Plains, Georgia, gave Billy a forum as they searched for stories on his much more serious brother. The entertaining ways of the president's *redneck* First Brother captivated a nation but were short-lived. Billy's notoriety from the media spotlighting his drinking and money problems led to a couple of serious missteps with the IRS and the Libyan government.

Between the time Billy was a relatively unknown Southern redneck in 1976 and a disgraced president's brother, there existed a magical moment in the summer of 1977. Two radio guys slipped into the tiny land of peanuts and Billy Beer, sampling the whole Carter peanut patchwork that gave us a *Twilight Zone* of memories.

I earned a week's vacation from WYEN and planned a road trip to Florida with *a side trip to Plains* joined by a close friend just as interested in radio as me. This may sound strange, but I figured we could talk about stories we covered, especially crime. It's not unusual for radio news people to retell crime stories. We'd discuss how the stories were covered. Mark Woolsey was among the best radio news anchors I'd ever heard. I valued his knowledge of news and handling of news stories. Not only was our summer vacation a chance to gain new experiences, but our vacation doubled as a news seminar on wheels.

Mark and I rolled into Plains. We had already vacationed in Florida and were starting our trip home on Route 75. On the way down to Florida, we thought about stopping in Plains first, but we had to check into a hotel in Fort Lauderdale by a certain time, so we saved the side trip on Route 280 to Plains for the return home. Our plans for Plains meant visiting everything Jimmy Carter, and with a little luck, we'd be doused in Billy too. With a quarter of a tank of gas left from driving through most of central Florida and parts of Miami and Fort Lauderdale, Mark and I now saw before us something we'd heard so much about—Billy's Gas Station with its real Plains character. I rode willingly into the world of Carter, but then I caught site of Billy's pump price.

"Eighty-five cents a gallon? That can't be right!"

"He can get away with it," Mark figured.

I figured filling up at Billy Carter's gas station was a uniquely American experience. In the relatively short period of time that Billy was our national "charming and playful" character, you could feel pretty good saying you filled up with Billy gas and talked to him.

Still, I'd been disappointed. I guess Billy can't be there all the time, pumping gas and telling stories. I mean gas is gas, but there's only one Billy, and he wasn't around.

Where else can you get someone entertaining you at their own gasoline station, saying something funny and embarrassing about you as a customer or about his older brother or about himself as a good old country boy?

"Maybe we can hold off and just find gas somewhere else in Plains," I reasoned. "What about going back to Americus?"

Mark seemed agreeable, but his tone hinted he wasn't budging.

"Did you see a gas station?"

"We could drive on this road here," I offered, searching the glove box for a map, then realizing Mark didn't rely on maps. All the directions from Chicago to Miami were in his head. I pointed straight ahead, without the hint of an idea where my finger might lead. Red dirt and peach trees, that's all I saw from Americus to Plains—maybe a gas station somewhere along the way. I don't recall, actually, since I wasn't really paying attention. I focused on this Florida orange from a bag in the back of the car. Anyway, we had barely enough time for Plains and the trip back to Chicago in one day. Mark had to drive on down to Missouri where he'd been news director of a radio station in St. Joseph. Doubling back to Americus would have taken too long.

Hard to figure how much time we could spend in Plains, but sitting around in the car deciding on whether Mark should buy Billy gas was wasting time we didn't have.

"We don't need that much gas," I said, bending slightly forward, pulling out my wallet, but Mark waved off my offer.

Mark rolled down his window. He saw someone from behind the car come over to the driver's side. The man leaned in toward the open window, looked at Mark, and asked, "Fill it up?"

We both instantly knew this wasn't Billy. No toothy Carter grin or welcoming hello.

"Yeah, go ahead," Mark responded with an almost imperceptible groan.

While we sat waiting for the attendant to finish filling the tank, I thought of Billy pumping gas and how the image would have stuck in my mind for years.

Several crates of Billy Beer were stacked near a door to the service station. We didn't notice the beer stock right away, but we'd heard Billy Beer tasted as good as sewer water. The beer with Carter's name was supposed to have been like his favorite Pabst Blue Ribbon.

"How about Billy Beer?" Mark asked, wondering whether a can was worth the price.

"We're not coming back to Plains," I figured, "so why don't you buy a beer?"

I convinced him to pay a buck for a can ... and why not. We had to know whether the *mystery liquid* was better than what we'd heard. Mark popped open a can.

"Wouldn't it be better cold?" I casually suggested.

"Probably." Mark shrugged.

He took a quick swig of Billy Beer and then became uncharacteristically quiet.

I looked at the can and gave in to my need to know.

"Would you mind if I tried some?" I put down the orange I'd been peeling, though my lap wasn't the best place.

Mark offered, "Here, have as much as you want."

I looked into the opening of the can, and it just looked dark and watery, no head or foam or anything. That's when Mark moved the car away from the pumps, and unfortunately Billy Beer spilled onto my shirt, but I also got a bit in my mouth, just enough to know whether what was said about Billy Beer being swill was true. I smiled at Mark. He knew and smiled back. He'd been a guinea pig, so now I'd be the same pig with swill. I had worse, I thought, like liver as a kid, but that thought didn't get the liquid down my throat any faster. I gave the can back to Mark, thinking he might drink some more, but he instead tossed the nearly full can in a garbage can at Billy's gas station.

We drove away from the station, turning from Church Street to Main Street, and that's when Mark saw two people just to the left of the town's shops.

"Stew, that looks like Billy Carter!"

"I can't tell," I admitted. Mark's eyes were apparently better than mine. Maybe Mark knew by Billy's distinctive barrel chest.

"We've got to get over there." Mark's reporter instincts kicked in.

I nodded, not as excited on this celebrity hunt, mainly because I didn't want to bother them. Mark had none of my anxiety. He found a spot between the shops and where the couple stood. I looked around at the little stores, all in a row, and didn't see a single person, tourist or native, walking around.

"Billy's right there! I've got to go," Mark persisted, pointing at the windshield.

From the front passenger side, I squinted from the glare off the hood of Mark's car. I didn't quite hear him, but the way he was getting out of the car, I knew he'd seen Billy.

He'd already taken five steps and left the car door wide open. To anyone else, this may have looked like Mark flinging himself headlong into stuff, but where some guys move without thinking, Mark analyzed things lightning fast, before he pushed forward. He briefly turned around. I waved him on. Now he was kind of running or maybe walking fast, but he wouldn't let the couple slip away. All I could see in the sun was this golden-hued backdrop of a building and two silhouettes. I assumed Mark met up with Billy.

With a little time to kill waiting for him, I reached into the back of the car and grabbed another orange from an open sack. We'd planned on splitting the other orange sack equally. So now, I smell citrus in the car while Mark is standing near the Carter Peanut Warehouse that Billy had most recently been operating. I'd been wiping my hands of fresh orange, but without a napkin, and the bottom of my shirt got a bit sticky.

I don't know what they'd been saying, but Billy turned to face Mark, and the woman gave Mark her attention too. All three stood within inches of each other, talking and talking. I felt like I had to join them or lose the chance to say I'd spoken to Billy. Mark saw me moving slowly toward him, and he smiled and waved slightly with his hand, encouraging me to join in.

"Where are you guys from?" Billy asked.

Mark answered, "I'm from Belleville, Illinois, not too far from St. Louis."

"I'm from Chicago," I added to Mark's quick response, thinking Billy wouldn't know Morton Grove.

"What are you doing out here?" Billy didn't sound redneck, whatever that's supposed to sound like.

"We had to see Plains," Mark explained.

Carter seemed satisfied. He just stared for a moment at us and then looked at his wife, and she smiled. Then he wished us well on our trip home. Billy disengaged from our conversation and turned away. The woman we

THE WYEN EXPERIENCE

took for his wife seemed not as ready to join him. She had a question or two of her own for us, and we answered. Then she too left, catching up to Billy. They headed into one of the factory buildings that dwarfed us and lent an uncomfortable balance between the downtown antique shops and the high-tech peanut factories.

For a few moments, Mark and I stood in the middle of their downtown, where our hosts had been.

"I think Billy actually had a red neck," I chided.

"I wasn't really expecting them to spend much time talking." Mark sounded surprised.

"How about we head over to the shops and see what's there before we leave?" I suggested.

"All right," Mark answered, but his sense of news again kicked in, and he said we must look for one particular place before we leave.

"Got to see Plains Baptist Church where President Carter worships."

"We could go and look for the church, but let's first check out at least a couple of the shops," I directed.

We walked toward the stores.

I discovered real peanut butter at one of the shops. This wasn't the stuff in the grocery stores that in comparison I imagine might taste similar to eating chemicals.

"What'd you get?" Mark asked. He saw me leave the store and waited outside.

I had two bags, one with a Peanut Patch ashtray, and the other was Peanut Patch peanut butter. The tray was my Plains souvenir and had the words *Plains, GA, Home of Jimmy Carter* printed on it, and an artist named JR Legg 1976 had drawn a big peanut body with a face and hair in the middle of the ashtray. The peanut man couldn't pass for Billy because it didn't wear glasses and most certainly wasn't President Carter because Peanut Man kept his mouth closed. The other bag with Peanut Patch peanut butter from Plains, Georgia, was kind of my experiment with real peanut butter. The experiment went well. This was sweet peanut butter, free of chemicals, and I couldn't get enough on my fingers. Yet, I felt restraint, maybe paper thin, but I'd hold back from finishing the whole jar. Maybe Mark wanted some?

"Mark, I'll save the rest for later."

I closed the lid, screwing it tight with an extra twist, trying hard not to use my fingers encrusted with peanut butter. I had worked two fingers into the peanut butter as a makeshift spoon. Mark watched this and said nothing. Maybe my double finger dipping was too out there for Mark? He had to already know my willpower with this candy-tasting peanut butter was exactly zero.

I just had one question for Mark.

"Where can I get a glass of water?"

Woolsey currently anchors news on weeknights on Atlanta's All News 106.7. Mark had worked as a radio broadcast meteorologist at The Weather Channel from the greater Atlanta area and as a reporter, anchor, and news director at radio stations in Illinois, Missouri, Kansas, Oklahoma, and Texas.

Chapter Twenty-One

GOING ABOUT THEIR BUSINESS

In the Family

"You look familiar," I said, conceding a hint may help.

Sometimes people popped into the WZSR-FM newsroom in Crystal Lake, but they usually were clients on tour of the station with an account executive. I'd been working on public service announcements. Now I'm staring at this guy, and he's smiling at me. He's wearing a white shirt and tie and black pants, clothes that indicate he's probably someone that came in to present a good image.

"Stew, we worked together at WYEN Radio," he offered, trying to bring me around to recognizing him.

I was just relieved he wasn't there to complain about something I said on-air ... so I didn't exactly hear him.

"What?"

"We worked together at WYEN. Does that help?"

I quickly added the years in my head. "That must have been thirty-four years ago."

"Yeah, I know I've changed a bit."

He was short like me. He had a deep voice, distinctive in its medium to slow delivery of words, very crisp and clear. He was a broadcaster or maybe a prison guard.

"Wait … you're Kenn Harris."

"Well, Kenn Heinlein, but yes, I was Harris on WYEN."

"You haven't changed," I insisted, but I lied. We had both gained some extra padding around the middle and more facial hair, though he wasn't as follicly challenged, but he probably could have used a comb right there. I didn't happen to own one.

"Stew, I had heard you were working in Crystal Lake, and I knew I was coming here to begin working in the sales department, so I had to stop in the newsroom."

Public service had to wait. I had to know more.

"I'm starting with the station's Bridal Showcase and working in the sales department."

We started reliving our WYEN days. He'd say something, and then I'd chip in, and we'd go back and forth. WYEN was always at the center of our little stories. We were either like two teenage girls who'd not seen each other in years and managed to stop texting for a moment and had so much catching up to do, or we were like two old men who hadn't seen each other for years, didn't know how to text, and thought with conviction the other had passed away.

"You married into the Walters family?"

"Jackie is my wife, and of course, Carol is my mother-in-law."

I knew both Jackie and Carol, but then this was more than thirty years ago, and a lot had changed, and I wanted to know about them and about what happened to the radio station, but most of all …

"Kenn, about fourteen years ago, Mr. Walters walked into my newsroom in Crystal Lake. The owner, Jim Hooker, took Ed on tour of the radio station, and I had so much to say to him, but Walters was pressed for time, and we never had a heart to heart that I needed to close the door on the WYEN chapter of my life."

"Stew, Ed passed away in 1999," Kenn said. "You'll never have that talk, unfortunately. I think he'd have enjoyed talking radio. He lived to talk about radio."

"I'm sorry, Kenn."

A half hour went by. I happened to look at my watch between WYEN stories. Then I let go in the air the words that I couldn't take back.

"We should consider writing a book on WYEN," I offered.

He looked at me and smiled, "A book? I know Carol could provide pictures, and I think she and Jackie would be clearly excited."

Without thinking through this much further, I sealed the deal, "Then let's do a book on WYEN."

"All right!" Kenn agreed.

THE WYEN EXPERIENCE

We were now on board to produce *The WYEN Experience*. Circumstances, though, took Kenn away from the writing part and more toward using his expertise on the photography. I accepted the challenge of writing the story on WYEN. For his part in the book, I asked Kenn to write his WYEN story.

I started working at WYEN in February 1978. Program Director Wayne Allen hired me to do the all-night show. Once I adjusted to the hours, it was a lot of fun. Unlike other day parts, I always felt a real connection with my listeners: those working third shift at area factories or the workers at O'Hare calling in every night. You could almost visualize them going about their jobs with the radio blasting away to keep them company. After a while, other opportunities opened up at the station, and I gave up the all-night show to become music director and host of Community Insight, the weekend public affairs program we taped during the week. This also allowed me the schedule flexibility to do an air shift on Saturday and Sunday mornings. Like the all-night show, those weekend morning shifts always had a loyal audience. Maybe not as large as the weekday audience, but they were out there, and every Saturday morning they'd call in with their requests and jam the phones for a chance to play "Morning Trivia."

There were a lot of great things about WYEN in the late seventies and early eighties. Granted the signal had its limitations, especially when it came to reaching Chicago and the south suburbs, but we owned the northwest suburbs. You could walk through the shopping centers at Randhurst and Woodfield Mall, and you'd hear us played on the radios in almost every store. With a heavy saturation of billboards and king-size ads on the suburban bus lines, you couldn't miss us. We even made it into the Academy Award-winning movie Ordinary People. *In a scene from Robert Redford's film, a Nortran bus fills the screen, and there on the side of the bus is a marquee with "WYEN 107FM" in blue and white.*

WYEN'S real assets were the jocks. They made the station come alive. For most of us, 'YEN was our second or third radio job. We had a good working knowledge of the biz, and now, we got to apply that drive and desire on a 50,000-watt suburban FM outlet.

One of the many great guys I worked with was the late Nick Farella. Nick followed me on Saturday mornings. Nick's family owned a bakery in Waukegan, and every Saturday morning, Nick would bring a box of fresh donuts to work, and we'd polish them off in no time at all. A fresh pot of coffee, a box of donuts, and a pack of Marlboros were all you needed for a four-hour air shift. Nick was one of the many jocks I had the good fortune to work with at WYEN. As our air shifts changed from time to time, Mike Roberts often followed me, or I would follow him. When my shift ended, I'd find myself sitting in the studio while Mike was on-the-air, and we'd talk for hours on end between songs. Radio was never like your average nine-to-five job. Unless you really had somewhere to be after your shift, it wasn't uncommon just to hang out at the station. It was fun to mess around in the production studio, or just hang out in the air studio with the jock that followed your shift.

ED WALTERS TOOK GREAT PRIDE IN THE PLACEMENT OF THE RADIO STATION CALL LETTERS IN THE OSCAR WINNING MOVIE, ORDINARY PEOPLE.

I never gave up my love of being on-the-air, and just doing weekends suited me just fine, but other opportunities existed at the station. WYEN also published a small entertainment magazine called Outside the Loop. With my background in art and graphic design, I saw a real challenge in OTL. I convinced Ed to give me a shot at running the magazine, so he turned the magazine over to me after releasing the woman who had owned the magazine. Now, on top of hosting Community Insight and jocking on the weekends, I was editor and art director of Outside the Loop magazine. Ed's daughter, Jackie, who would go on to become my wife, worked with me on every aspect of the magazine, including the monthly grind of heading out to the printer in Elgin, loading pallets full of magazines into the station van, and driving all over the northwest suburbs as we delivered the magazine to area businesses. But like everything else with Walt-West Enterprises, you wore a lot of different hats, and in this case, having a paper route was just another part of the job.

I left WYEN as a full-time employee in 1984 but kept my weekend shift. And even after Ed sold the station, I stayed on through two format changes and an owner I swear had fewer brains than a pet rock. I called it quits in 1989. The

WYEN that was so much a part of my life for so many years didn't exist anymore, and the people I had so much fun working with were long gone, so leaving was easy and long overdue.

Overnights with Mr. Roberts

Kenn had his favorite radio announcers as a kid. Just about all of the WYEN talent can remember a broadcaster they idolized. In the Chicago area, the name Larry Lujack on WLS AM was magical. Mention the name to me or anyone that never got enough of Superjock, and you might as well have waved a magic wand over our head. We return to the 1960s and early 1970s; we're a kid again with a transistor radio, and Lujack is telling us about Animal Stories. I had a similar reaction for another announcer during the time I attended Maine East High School in Park Ridge and worked as a busboy at Oscars Fine Food in Morton Grove. Larry *the Legend* Johnson performed his magic overnight on WIND-AM from 1970 to 1973, a short time in Chicago radio, but he got tired of overnights. Can't say I tired of overnights. Johnson kept me company some nights after I came home from working at the restaurant. I'd tune to WIND and hear his familiar Southern accent and thoughtful insights. He was a talker. My parents were asleep, and the house was quiet, except in the bathroom, where I had the fan on high and The Legend on higher. My clothes smelled of a mixture of beef, baked bread, steaming French onion soup, hot coffee, and cigarette butts in ashtrays. The grease settled in my hair, plastered down from putting my hands through it after carefully sliding butter patties into a bowl with ice for the newly turned over tables. The bathroom is where Johnson captured my attention. I plugged in a small radio and set it down on the toilet seat cover and stepped into the shower, scrubbing for a good half hour all eight hours of restaurant smells from my pores and hair, and Johnson's voice kept me awake in the shower.

Roberts has never tired of overnights, nor have his considerable audiences tired of his sound, decidedly upbeat and youthful. *Overnights with Mr. Roberts* are a different kind of enjoyable from the years of The Legend, but Mike's place in Chicago radio history is set for his longevity on-air. After WYEN, Mike moved on to what was then WCLR in 1982, which later became WTMX (101.9 THE MIX). For the majority of the past twenty-eight years, Mike has worked overnights, midnight to 5:30 a.m.

I asked Mike to write his own story of his time at WYEN, starting in late 1977.

My first shift was on New Year's Eve. I believe it was the 6 p.m. to 10 p.m. shift. Nick Farella followed from 10 p.m. to 2 a.m. I was twenty-one at the time.

For about the first year, I was part-time. Later in 1978, I became full-time and worked afternoon drive for a few months, then overnights until I left in early 1982.

Both Greg Allen and Nick Farella were friends of mine who I admired and respected for their talent and knowledge of the radio industry. I worked with Greg in Dundee, Illinois, and he helped me to get the job at 'YEN. Greg was already working there. He convinced me to put together an aircheck, and he brought it to Wayne Allen. A short time later, I interviewed with Wayne and got the job. Farella was interesting to talk to because of his experience in many aspects of radio throughout his career. Even after we both left 'YEN, Nick would always give me a call now and then to see how things were going at work and with my family. He always seemed so passionate about all aspects of radio (from management to on-air to promotions). I was saddened to hear of the passing of Greg and Nick. While I wasn't able to attend Greg's funeral, I did attend Nick's, which was mostly a gathering of family and friends, some who traveled from out of town to be there. There were people in the business I hadn't seen in years. There were people in radio who I knew of but had never met. It was interesting to hear how their paths crossed with Nick throughout the years. And those years do pass quickly. It's been over thirty years since I left WYEN.

My approach to working overnights hasn't really changed too much. While formats change, technology is probably the biggest change. As an example, WYEN in the 1970s had two turntables, three cart decks, reel-to-reel decks, and a splicing block and tape. I've also had the use of point-and-click and multiple screens and digital editing. It's amazing how those aspects like editing have changed. Sometimes, one of the biggest challenges of working the overnight shift is staying alert and awake if your sleep was compromised somehow during the day. On any shift, I found another challenge in locating a lot of the music from song requests that I would receive, especially when I first arrived at WYEN. There was no scrolling through a database of songs. It was more like walking into a hallway lined with rows and rows of albums (you know, 33 1/3 LPs). Back then, we'd get a lot of song requests for what I called 'YEN songs (songs that were well known to the 'YEN listener, great album cuts, and not necessarily the huge hit from that album). It took some time finding those songs ... since some I had never heard of at the time. Even now, I'll sometimes go online and listen to some of those songs that I closely identify with during my time there, especially my first year, songs like "Las Vegas Turnaround" by Hall and Oates, "Songbird" by Jesse Colin Young and "Walking in Rhythm" by the Blackbyrds.

THE WYEN EXPERIENCE

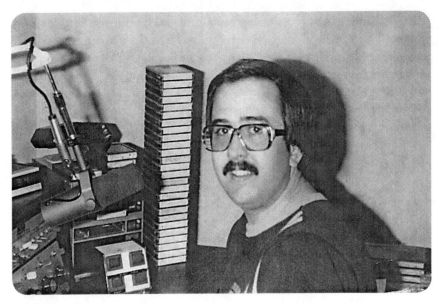

From the Walters Heinlein collection.

In my just over four years at WYEN, I worked and became friends with some very talented people: Kenn Harris (Kenn Heinlein), Wayne Bryman (Roger Leyden), Mark Dixon, Louie Parrott, Wally Gullick (a college friend and coworker), Jeff Dale (Puffer), Beth Kaye (Krusich), Wayne (Alan Weintraub), Nick and Greg, and of course Stew Cohen, just to name a few.

Every now and then, I'd see Walters and have a conversation, though today I can't recall the substance of those conversations. Working the hours that I did through most of my tenure at WYEN, most of my conversations were with the air talent who were on directly before and after me. I'm grateful that Ed gave me the opportunity to work there.

I'll always remember the details of WYEN that made the station what it was for both the announcers and many of our listeners, things like Request Radio ... the phone number 591-1166 ... some great (long) jingles ... "Now, more of your kind of music" ... even when you heard the weather, "dry weather chances" ... and also "WYEN Metro News/Weather" ... and "Early Morning Trivia." I also can't forget the butterscotch candy in a jar by the front desk. And within those four years, I graduated from college, got married, and we welcomed our first son into the world.

From the multitude of song requests from listeners to the people that I had the pleasure to work with, and to those who were responsible for me being at WYEN ... thank you.

Kenn and Mike are very much aware of the importance of the voice in radio. To say voices are everything is not overstating reality in broadcasting. Content of course will keep listeners, however, an annoying voice is very uncomfortable and probably won't generate loyalty unless the content is unique. In the next chapter, one of the most successful WYEN announcers had such an incredible voice that he was destined for a national radio show. Also in the next chapter, a man whose voice set him apart from all voiceover talent in Chicago found a home at WYEN.

Chapter Twenty-Two
SMOOTH OPERATORS

Next Stop

The Saturday night bars of seventies Chicago were living and breathing live music. The drinks were on the rocks, and beautiful young women in skirts filled the minds of guys crowding two and three deep behind bar stools.

Two young suburban guys wearing bell-bottom pants and flowery rayon shirts with two-tone platforms got into their car, a mint green Century Regal with a white top. Jeff kept his car in mint condition—so well shined you'd see your teeth on the hood if you leaned over far enough. He and I drove downtown. On the radio was Jimmy Buffet's "Margaritaville," but this was Saturday night, not a night for WLS or WCFL, but for a guy on radio that had a way of touching our souls, of programming music to lift us, separate us from our troubles. His name was Dick Bartley. On WFYR-FM, Chicago, a show we knew as original Saturday Night Oldies, Bartley's musical selections set the tone of our evening, easing us into the mood for dancing and conversation. The music of Saturday Night Oldies took the edge off loneliness. More than anything else in those mid-1970s of bar hopping and hot live music, Jeff Brenner and I remember just how many lonely people filled those bars. So we turned to Bartley and to Fred Winston, another WFYR announcer, and they helped ease us through the 1970s with humor and great music. Just as Bartley replaced an announcer on WFYR, **Bob Worthington** replaced Bartley's syndicated show in New York, yet each host continued playing oldies. Many a Saturday night, the music resonated deeply, our steady friend, feeding our souls with Blondie's "Heart of Glass," Billy Joel's "Only the Good Die Young," Franki Valli's "Grease," and ABBA's "Dancing Queen."

I'm proud to know Worthington, one of the successors of the great oldies show that Bartley broadcast coast-to-coast. Worthington had his turn in national syndication after leaving WYEN for New York, and he's worked at some of the biggest radio stations in America.

Bob possesses one of the most "I can listen to him all day" voices in radio's long, distinguished history. To describe his voice, you might think of Frank Sinatra's smooth phrasing. Sinatra's music fills the air, and Worthington's voice, whether on commercial spots or introducing songs, has a way of making people feel good too. Bartley performed similarly, and each host peeled away layers of the stuff that blocked full appreciation for the music. For *The WYEN Experience*, I asked Bob to tell his story.

Through my time at WYEN Request Radio, January 1983 to January 1986, as the midday and later afternoon jock and production director, I have mostly very good memories of Ed and Carol Walters, their son, Mike, and daughter, Jackie, and staff.

I know just about every deejay who wanted to make it downtown hoped to work on one of two stations in the suburbs: either WYEN in Des Plaines or WAUR in Aurora. Both were 50,000 watts and covered most of the metro area. I was hired from WCGO/Chicago Heights and WTAS-FM in Crete after working there while in college, and WYEN to me was a jump in the ranks!

I think of WYEN as a miracle in my life as I was able to do some part-time at the old WCLR while I was at WYEN, and in May of 1986, I made the huge jump from WYEN to the nation's number-one market at Mix 105 WNSR in New York City. While still working afternoons at WNSR, within two years I was host and producer of Solid Gold Saturday Night for almost nine years.

Ed mentioned to me he thought in the history of the station, many jocks went to downtown Chicago stations, one or two went to Los Angeles straight from WYEN, but he said that I was the first jock to go straight to New York City from WYEN. I know I was extremely lucky, and it was nothing short of a miracle from God. All said, WYEN was pivotal in my career, and I think of the staff, the 2400 Devon Building studios, and can't help but glance from the tollway on the way to a voice gig and see the building from I-294 that once housed the station, a lot of great memories of when we were all a lot younger.

THE WYEN EXPERIENCE

From the Hamilton Hotel in Itasca, it's hour number twenty-one without sleep for WYEN announcer Bob Worthington. He reminds his listeners that the Jerry Lewis MDA Telethon needs their support on Labor Day Weekend of 1984.

I recall some of the advertising clients on WYEN. As I mentioned earlier, I was the production director. Cinderella Rockefella, Knickers Lounge, Schaumburg Ford, Old Volo, Interiors By Bruce, The Black Ram Restaurant, Just Jeans, Old Style (Remember the WYEN Old Style Mug Nights?) ... I still have a few WYEN mugs around! Some other advertisers I remember: Randhurst Mall, Bredeman Buick, and Morrie Mages. WYEN carried the Chicago Sting one year. I did an appearance with Sting player Rudy Glenn in the Golf Mill parking lot one hot summer day. One promotion that elevated us into the book with a .05 share was the WYEN "Stick with Us and Win" bumper sticker promotion. I remember voicing the promos—fun times at WYEN!

Worthington and the next former WYEN announcer and voiceover talent have the voice quality that stops you. They could simply say, "How are you," and it's the most incredibly wonderful sounding *how are you*. But before you learn about our second golden voice, maybe there is an overemphasis on the significance of the voice. That's why *The WYEN Experience* calls attention to someone whose voice can literally send shivers down one's spine.

Best Voice

The man with "The Voice of God" described professional football between the Dallas Cowboys "Doomsday Defense" and Pittsburgh Steelers "Steel Curtain" in the 1970s NFL films. He used the word *Armageddon,* and his legend took off. I imagined he used Armageddon with the Green Bay Packers, the Chicago Bears, Steelers, Cowboys, and other teams and mixed in the conditions.

He'd thunder, "It's Armageddon under a blue sky." Maybe he'd change it up, "The frozen tundra of Lambeau Field portends Armageddon."

John Facenda had an unmistakable voice. His timbre was so powerful that had he been a battlefield general, soldiers would gladly have walked through a minefield for him. Facenda narrated NFL Films and Football Follies with a voice that legendary sports announcer Howard Cosell said could make the coin toss sound like the world was ending. Using his powerful delivery, Facenda created drama and poetry, truly influencing my style of reading news on radio. Though very few news stories on WYEN were treated as Armageddon, the Chicago winter weather in the late 1970s sure seemed to present itself as such.

"A foot of snow has descended from the heavens, burying our cars, covering streets and sidewalks. The trip for a loaf of bread and milk has become an adventure only the brave can endure."

No, I never would dare string together phrases like "descended from the heavens"—just too much drama for a WYEN Metro Newscast. This isn't in my character to describe weather as though the Ice Age had just returned to Des Plaines. I could imagine Facenda or Walter Cronkite saying, "On this bright and crisp day, Armageddon has come to Des Plaines in a most horrible way with a foot of snow and a temperature of twenty-five below." I did not know Facenda personally, but if he were alive today, I'd probably form a club with honorary names such as his, Cronkite, James Earl Jones, and Morgan Freeman and call it The Armageddon Voice Club. When these men said something, you believed it. I know of one more member worthy of immediate entrance into this exclusive club—**Jack Stockton**. His is not a name familiar on television or radio. Yet Stockton's voice was so powerful and commanding on many radio spots on WYEN, he brought a special sound to suburban radio. You won't hear his voice on-air today. But I'll try describing what you are missing as though his voice were a deep-dish pizza. So rich, Jack's voice packed layers of cheese, mozzarella and provolone. That golden brown crust topped with thick sauce over the melting cheese was Jack's delivery, just the right tempo for a full feeling. Mixed in the sauce, you'd discover the freshest spinach and mushrooms, giving us Jack's distinctive voice that only he could spread among his listeners. Aside from the food references, if Jack insisted

THE WYEN EXPERIENCE

on-air I wear a certain winter coat, I'd wear that winter coat, and should he suggest I fly to Hawaii for vacation, then I'd plan my Hawaiian vacation with my travel agent, and if he said I should eat Spam, well, I'm not that stupid!

WYEN had a sales staff that said their client wanted Jack to do the spot and only Jack. His voice made WYEN sound instantly competitive with WLS and WCFL. Hearing his commercials, I'd feel that WYEN was truly a station that belonged in this top market.

Stockton said, "I'd get off the air at WCFL and get into WYEN at 2 a.m. and produce spot after spot, such as Woodfield Ford and Old Volo Village. Ed would pay me five dollars for each commercial I'd do. That about covered the price I paid for gasoline to get to WYEN. Funny, how Ed was a bit apprehensive in the beginning to hire me, but agreed on minimal pay and on the night shift that I asked. This freed me to do more voice work in commercial and industrial. Eventually WCFL News Director Jim Frank heard me and hired me on the beautiful music station. Ed came to me and asked me to continue coming out to WYEN to cut commercials. I needed permission from WCFL and got it. The station said okay, so I came out."

Today, Stockton works with his wife on their travel agency and spends time visiting family members in the Sun Belt. His voice, I can vouch for, is still as strong as I remember all those years ago.

Greg, Paul, Chris, Dan, and Bob were companions of WYEN listeners. These announcers didn't just play music and introduce songs, they kept their listeners company. Live radio today continues offering a close bond between announcers and listeners similar to the WYEN of more than forty years ago. People are still making comments on issues of the day, asking for a song, or talking about their children or themselves. A strong bond allows for such discussion that is not possible with iPods, CDs, newspapers, or television.

Chapter Twenty-Three

THE EVERYMAN TO THE MARATHON MAN

Greg Stephens

He hung a movie poster of Olivia Newton John on the wall above his office desk in a room at a converted warehouse. This was not a typical poster you might find in a store. This gigantic poster equaled the size of Greg's crush on the Australian singer. No one at his office inside United Cable of Carpentersville seemed to mind Newton John staring them down every time they came into Greg's office. All that mattered was Stephens was content. Another movie poster would arrive the following month and the month after, but he had the one he wanted.

My "Newton John" visit with Greg at his office was the last time I'd see him. A few years later, I'd learn of Greg's passing. Greg and I hadn't seen each other because our schedules tightened, but nonetheless, I'm sure independently we thought about getting together. It just never occurred that time could run out.

Long before Greg left radio for cable television management, we worked together at WYEN, shared Friday night dinner conversations at a popular restaurant, and attended parties with friends.

Coincidentally, Greg and I left WYEN within months of each other. I left in early 1979 and moved away from the north suburbs for the northwest suburbs to work at WIVS-AM and Mal Bellairs and his family. Greg didn't relocate. He lived in northern Kane County and now didn't have much of a commute to United Cable. I'd find a place quickly in Algonquin, a little studio inside a house next to Vern's Taxidermy on Main Street, where the painted ladies stood side by side with hardly a space between. Most weekends, I called Greg, and we gathered with his friends at the Village Squire in Carpentersville or at one of their homes. With them in 1979 and 1980, I was simply Stew, and Greg was either Stephens or Allen, but not the alternative, Wojewoda, his real last name.

I first heard Greg on my car radio one night. I'd been heading home, and my speakers were blasting WYEN's jingle, greeting Greg Stephens. I didn't turn up the volume, but I listened, waiting for his voice to fill my car. The song ended, and a pleasant-sounding nasally voice made its way through the driver's side and into the back of the car. The "nasally voice" had a personality and a way of sounding like everyman. Greg didn't try to do any vocal tricks or sound deeper than he was. I expected this guy to have worked as a postal carrier or a youth baseball coach. The diction was not perfect, but the energy was solid, and the feelings were true.

FROM THE WALTERS HEINLEIN COLLECTION.

Former longtime WCFL Production Manager Tom Konard, a friend of Greg's, taped Greg's early shows before he began working at WYEN. Konard sent me a tape from his home production studio where he lives in Belgium.

Bill Emerson and Greg Allen on WVFV Dundee, Illinois, on April 23, 1976

Bill: I always say you are cute and loveable and cuddly because that's what everybody tells me on the phone here. They'll call up and ask where is the cute, cuddly, and innocent Greg Allen? I've got a couple of prerecorded jock intros here.

"Greggy Pooh Allen on WVFV 104."

Greg: I know you didn't have to pay much for that! (laughs).

Bill: Listen, on my budget?

Greg: Your budget?

Bill: Here's another.

"Greggy Pooh Allen." (sings ... rising to Allen.)

(Greg laughs)

Bill: You haven't got much to say.

Greg: No, I ... Is it a weather jingle coming up?

Bill: No, here's the last one, and it's the big production ... so you get the full impact, we'll do it over a song. I'm going to show you exactly how it works. Ready? (softly says) Greg Allen's personal jingle.

(sexy whisper) "You are listening to tall, dark, and handsome Greg Allen."

Greg: (softly) I've got my own jingles, man. I feel great on a Saturday night at 8:28.

(music ... *This is for all the lonely people, thinking life has passed them by* ...)

Terry Flynn, a close friend of Greg's, called me one morning at work and told of Greg's passing.
Southern nights ... have you ever felt southern nights, free as a breeze ...
WYEN, The Station That Sounds like You've Made It.
"Greg Stephens at 10:09. Good Monday night to you on Request Radio at 591-1166. Here's a song for Elgin tonight."
Greg settled back in his chair for a few minutes listening to a favorite performer, and then he got up and found the next few records from listener requests.
"Here's Paul Davis and 'I Go Crazy.' Did you have a crazy weekend?"

Greg didn't expect callers to tell him what kind of a crazy weekend they had, but if they called, he was ready to talk and maybe air the craziest weekend.

"'Hit the Road, Baby.' That's Dionne Warwick, request music for Hanover Park at 591-1166. Here are the Turtles from a few moons back on WYEN, the station that sounds like you've made it."

Konard's tape ran out. That's all I had. For all the thousands of songs he introduced and all the jokes and weather reports, my memory of Greg is mainly of us sitting at the Garlic Press, a little fancy restaurant near the radio station. We drank beer, occasionally Sangria, and usually ate something with lots of noodles, and by the time I walked away from the table, I was refreshed and ready for another week of work. That's how I choose to remember Greg, sitting with me at the Garlic Press, talking radio.

Paul Brian

Although he wasn't *the driving force* behind WYEN, WGN 720, or WLS AM, Brian mapped the route for his career without the need of GPS and with stops along the way in Des Plaines and Chicago. Paul worked as an announcer at WYEN and WGN and eventually *sat in the driver's seat*, hosting the WLS radio show Drive Chicago and *put the pedal to the metal* once he'd been named director of communications and spokesman for the Chicago Automobile Trade Association, CATA, and *he never had to take a backseat to anyone.*

Clichés aside, researching Paul before my interview with him for *The WYEN Experience*, I checked the WLS 890AM website and found he's the broadcaster and analyst of Drive Chicago on ABC Radio Chicago. In one picture of him behind the wheel, Brian is sitting in a teal-colored luxury car with light brown leather seats. Paul is described in the website as "one of our nation's best known and respected automotive industry experts." For a number of men, Paul has combined two of their favorite things in life, music and cars. Add a third, food, as Paul is described as an accomplished chef and winner of the 1983 World Championship Chili Cook-off. But in his position as director of communications for CATA, Brian is very much involved in producing the annual Auto Show at McCormick Place every February. He was also the primary fill in for WGN radio legends Wally Phillips and Bob Collins.

Before all of these things, Paul started at WYEN.

"I would not have had a WCLR FM or a WGN AM or anything else in my radio career if it weren't for WYEN. Everybody's got a first step."

The US Army veteran's first step had him serving American Forces Radio and TV in the Panama Canal Zone.

"WYEN was my transition job out of the military into civilian life. That's what interested Ed Walters the most. The army had put together a program called Project Transition. If I could find a civilian job in the last ninety days of my enlistment, and whether I stayed or not, the army would pay for the internship with the company. Many of them turned into full-time civilian jobs."

The last military assignment brought Brian closer to home during his stay at Fort Sheridan. From there, he contacted Walters and explained how Project Transition worked.

"I suggested Ed could have me all week long, and the army would still pick up my tab ... and I have to say this appealed to him a lot. Yet I remember Ed's decision going down right to the last day or two before my discharge, and I finally put it to him whether I had a job ... on Monday or not?"

On the last day possible for a decision, Ed said Paul had landed a job at WYEN.

His memories are of coworkers Rob Reynolds, Dave Alpert, John Zur, Garry Meier, Ray Smithers, Bruce Elliott, and Gil Peters, but his most interesting story is of someone at WYEN whose impact came after his time at the station.

Alpert received permission and hired George Adams (real name George Adamkavitch) to do morning news. Brian says Adamkavitch went on to become the voice of Cap'n Crunch and voiced a lot of cartoons.

"Well, George often asked me to play a song for his mother, Ramona. So after he red dogged me to play the song 'Ramona' every couple of weeks, I would for his mom."

Brian tells a story I had heard from several former WYEN announcers in researching for the book. Paul became one of a handful of announcers who said WYEN stood for *Where You Earn Nothing*. Brian was responding on-air to a listener who wanted to know whether YEN was somehow connected to Japanese ownership.

"That wasn't the end of my career at WYEN, but shortly thereafter, so it's always stuck in my mind that saying WYEN stood for Where You Earn Nothing was one of the feathers hopping on the camel's back."

From WYEN, Paul says he sold cars for Dan Wolf Pontiac in LaGrange and eventually landed work as an afternoon announcer at WCLR-FM. He was close friends of the station's Program Director Jack Kelly. Kelly hired Brian to do afternoons, but then Bonneville bought KAFM-Dallas, Texas, and Brian went there as program director. He eventually switched to WFAA-Dallas and worked as program director and afternoon drive, and came home to Chicago and WGN 720AM, where he worked from 1984 to 1989.

"At WGN, I was filling in for Wally Phillips and Bob Collins. My regular slot was 6:30 p.m. to 9 p.m., when Milt Rosenberg came on. When Wally or Bob would go on vacation, management first had me fill in for Bob in the afternoon, and Bob would fill in for Wally in the morning, but then management decided to have me fill in directly for Wally and let Bob stay in the afternoon."

Paul remembers the first time he walked into what was then called the Pierre Andre Memorial Studio.

"Jack Brickhouse was on-the-air doing Sports Central. He says, 'Hi, kid, come in. How are you doing?'"

Brickhouse was textbook Jack, according to Brian, "So I walk into this place which is like looking at the burning bush, walking into the Holy of Holies, and there's Brickhouse!"

Paul thought the legendary sports broadcaster would deliver something to him he'd never forget, and he did, but not exactly pearls of wisdom passed down. Instead, Brickhouse injects, "I'll be with you in just a second."

Brian was part of a broadcasting family with Collins, Harry Caray, Brickhouse, Roy Leonard, Phillips, and Eddy Schwartz at WGN. From Dallas, WGN's Dan Fabian would fly Brian to Chicago to fill in for Schwartz overnight. So his decision to leave radio did not just happen one night. Considerable thought preceded his move away from radio in 1989. He spent the next three years working for Alfa Romeo on the IndyCar team. Part of the time he worked for Alfa Romeo he spent living in Milan, Italy. From there, he helped form the IndyCar Radio Network with Paul Page from ABC, becoming the radio color guy for the IndyCar Series until he joined the Chicago Automobile Trade Association, as CATA's spokesman and director of communications to this day. CATA produces the Chicago Auto Show at McCormick Place.

Ray Baldy

He saw radio! Ray didn't just listen to the airwaves in the fifties and sixties, thanks to his father. His dad, Leonard Baldy, made sure by introducing Ray to WGN 720AM in Chicago. Ray's father became the first *Flying Officer* for WGN, and sometimes he'd bring his son to the WGN studios. A young Ray watched announcers Eddy Hubbard and Wally Phillips.

"I got the bug," Baldy said, "because I loved being around the radio station when my dad worked for WGN Radio."

Ray recounted the story of his father and pilot on the day they were killed in a helicopter crash in 1960. Officer Baldy had reported on traffic from the sky for two years. Ray was only nine years old. His sister was six, and his brother was two on that tragic day in May.

Ray forgot none of the details.

"When you were on the radio in a city like Chicago, you were as big as John Wayne was in Hollywood, and so my dad was extremely popular for his reports."

Even before the helicopter's top rotor blade came off while Baldy and his pilot were in flight, Ray knew his future was radio. He'd been immersed for so long, but on his dad's passing, the WGN visits ended, though Hubbard did come by and take him to a few ballgames. Ray got to sit in with Lou Boudreau and Vince Lloyd in the WGN Radio Cubs broadcast booth.

Another type of booth waited for him for his radio internship in the early 1970s. Ray had already met WYEN News Director John Watkins at Columbia College in Chicago. As a student, Ray interned at WYEN before he graduated. The internship turned into a part-time deal, and he began putting in ten to fifteen hours a week at WYEN, doing some news and traffic and working with Watkins in the morning.

"I'd do a little story, a *phoner* from my apartment. They'd *feed* me the story. I'd rewrite it and read it over the phone. We made it sound like we had a bunch of reporters even though we only had a few altogether."

In the two to three years Baldy worked at WYEN, he learned how to do voiceover work by listening to the national guys and how they delivered a spot. He'd write down the copy, word for word, and try to emulate what they were saying and how they delivered the message.

"I spent a lot of time in production, but mostly the production wasn't going anywhere other than to my own demo."

Baldy produced a pretty good demo. He learned so much by practicing the craft. Ray found his services in demand at a Top 40 station above a Laundromat in Ottawa, Illinois. WOLI Program Director Pat Martin thought something had to go, and for a year, Ray became Max Cooper, the deejay. Baldy also tried on Jack Fever and Jack Elliott. He moved to Phoenix on the sale of WOLI. During the time he was the morning WOLI deejay, he'd been making contact with KRIZ. The program director hired Ray, but he learned the guy he'd replace needed to stay an extra month. The KRIZ program director suggested Ray work weekends at KSGR, an oldies station. He knew the program director at KSGR needed a weekend announcer.

"I met with the program director at KSGR and agreed to work several weekend shifts over the next month, and I used the name Jack Elliott. I became Jack Fever a month later when I went on-air at KRIZ. The program director suggested going back to Jack Elliott, remembering the name from my short stint at KSGR."

Ray did nights and weekends at first at KRIZ, and then later the evening drive.

"I was only Jack Fever for a weekend or so at KRIZ before the program director made the name change. Glad he did; otherwise I would have been accused of stealing from WKRP, the TV show, and Dr. Johnny Fever."

Then he left Arizona on the sale of the station and went to Oklahoma City in 1978, working for the third oldest station in America and the first radio station west of the Mississippi River, WKY.

"WKY was like WLS and had a full fourteen-person news department, and we still played hits, and I stayed twelve years through a number of format changes until they flipped from country to beautiful music in 1990."

Sometimes the best things for a career happen at dinner, and that's exactly where Jack's career took off—at a dinner with the general manager and program director at KLTE K-Lite, which eventually changed call letters to KOQL FM in Oklahoma City, an oldies station. Ron Williams was handling news in the morning, and they had known each other since Jack came to Oklahoma City.

"The general manager and program director hooked us up on mornings because they were looking to make a change in the morning show. Ron and I have been together since 1990, first on KOQL and for the past seventeen years on KYIS FM in Oklahoma City."

From his internship in the early 1970s at WYEN in Des Plaines, Illinois, to Oklahoma City at 98.9 KYIS (KISS), Jack's road has given Oklahoma a top-quality broadcaster.

"I've been really fortunate. Oklahoma has been very good to us (Jack & Ron Show) in so much as we've won a bunch of awards. We are always conscious of the show's quality and content. We've managed to maintain a high level for a long time because the demographics and ratings have been strong all along."

The number and quality of the awards won by the KYIS Morning Show with Jack & Ron explain why they are a top broadcasting team in Oklahoma. Among the awards are two Oklahoma Association of Broadcasters Best Radio Personalities of the Year. The awards list goes on and on, highlighted by three times receiving the American Women in Radio & TV Award for Best Radio Personalities.

From Jack's internship and part-time work at WYEN to his position as half of the morning team of Jack & Ron, he has found a deep-seated motivation that's taken him from Des Plaines to Ottawa to Phoenix to Oklahoma City. His dad's tragic death in the crash of a WGN traffic helicopter opened a young Ray to the world of radio, but his desire to entertain and inform has kept him riding the radio waves to continued success and recognition.

Steve Kmetko

He saw Betty Davis eyes and talked to *Pretty Woman*. This was the role "I was born to play," Kmetko often said of his terrific job as an anchor for E! News Daily.

At the peak of his career, this Cleveland, Ohio, native reminded himself he had a great time as the anchor of newscasts for E! News Daily. Steve worked at E! on entertainment and traveled around the world visiting the Cannes Film Festival at least ten times, the Venice Film Festival, Berlin and Sundance Film Festival. Because of his association with E! and the CBS Los Angeles affiliate as an entertainment reporter, Steve came in contact with Julia Roberts, Robin Williams, Jody Foster, Sophia Loren, George Clooney, and Davis, and many other entertainers.

"You name it, I've interviewed them. I've also had the opportunity to interview presidents because I covered hard news as well."

Steve's career started in the early 1970s, long before Kmetko made the switch from radio to television and from the Midwest to the West Coast.

"I listened to traffic reports on WGN 720 and used a form in the WYEN news room with all the major intersections and thoroughfares."

For fun, I gave Steve an assignment. I told him to pretend he's back at WYEN Request Radio reading from the form he filled in with traffic problems he'd heard on WGN. Reprising his role of WYEN Metro Traffic Center Reporter, Steve's voice expressed urgency in his delivery.

"The Edens Expressway from Montrose to the junction is fifteen minutes. I would repeat like a parrot what I had just copied down. It was all kind of funny, but it was a good experience for a college kid, and that's what I did."

Though Kmetko's WYEN internship through Columbia College lasted less than a year, broadcasting had already become a vital part of his life.

"I learned about things like deadlines, showing up for work on time, listening carefully, writing fast, and getting up early and looking alive. I found out I wanted to keep doing it."

Steve often just hung around the WYEN studios, not doing anything but talking to the on-air staff, reading wire copy, and immersing himself in the setting.

To this day, Steve has maintained his friendship with Gil Peters, one of the first announcers on WYEN.

"Gil helped me get the internship. We'll never lose touch though we may not speak for a while."

In 1977, Steve went on vacation in Rhinelander, Wisconsin. He visited a friend there, and as he left to return to Columbia College and his WYEN internship, he stopped at the local TV station. His friend's parents had suggested he drop by because they believed the station would hire anybody, and sure enough they did.

"They literally sat me down in front of a camera, and I ripped copy off the wire and read it."

The general manager took Steve into his office, and the next thing Kmetko knew, he was returning in a week, starting a job on a television station in Rhinelander. From there, he worked in Green Bay, Grand Rapids, Louisville, and Los Angeles at the CBS-owned and operated station. All this moving in only five years! Kmetko worked at CBS for ten years, E! Cable for eight and a half years, and at Fox for three years on a freelance basis, and his acting credits include the movie *Zoolander*.

Pervis Spann

One night in the late 1970s, I walked into the studio where the All-Night Blues Man controlled the sound of WYEN. I was just leaving after a long evening of news writing. The on-air announcer shifts had changed, and the legend in blues sat down with his entourage of production assistants hovering behind him. I had to meet the Blues Man. I'd heard the deep voice and the blues out of the speaker in the newsroom. Spann turned in his chair, put out his hand and introduced himself. I already knew him by name, and he probably knew that as well, but for him to say his name to me alone … that was a memory I'd have forever. So few watched him; so many heard him. I was the former … Jack Johnson, a WYEN sales associate, was the latter.

Johnson mused, "My understanding was it wasn't unusual for him to be on the phone pitching advertisers while a record was playing and while he was also simultaneously loading the cart machines."

Spann's legend grew at WVON in Chicago. His influence as a broadcaster brought recognition to blues musicians, and his association with B.B. King helped further his reach in the world of blues.

Sometimes an announcer stayed for a very short time, but whatever the length, Spann and Greg Brown remembered their visit for their entire career.

Rob Sidney

Program director for 101.5 LITE FM Miami, Sidney wrote his memories of WYEN for *The WYEN Experience*.

My experience at Y107 was brief but memorable; in fact, it was my first full-time radio paycheck (all $200 a week), doing five overnights and a Saturday 8 p.m. to 1 a.m. shift in the summer of 1983. I was in between my junior and senior years at Northwestern University, taking a token summer class to justify living in the dorms and trying to fit in work as operations manager at the campus station, WNUR. No car, so I'd have to take the El train from Evanston to the Washington station downtown, then out the northwest line to the River Road station and hoof it the rest of the way to the WYEN studios.

I took over from Jeff (Puffer) Dale at 1 a.m. and was relieved by Kevin Jay at 6 a.m. Exhausted, I'd nap for about ninety minutes on the lobby couch and then hang out to write copy, dub spots, and chew the fat. I learned how to wind a cart at WYEN (no radio station I'd worked at before or since had been so frugal as to reload its own carts) ... and how to backtime into network news (always a challenge, since it wasn't like there was a synchronized master clock system or anything).

The programming was a hodgepodge of Adult Contemporary and Oldies and the occasional jock's-choice album cut (played literally from records). I recall being told consultant Paul Christy had set up the rotation system (which included current/recurrent categories designated by letters A through E ... and non-currents categorized by number); later, it had been bastardized somewhat, which would account for the curiously absent "B" category.

Ed and Carol Walters were always kind to me ... daughter, Jackie, was a sweetheart ... and son, Mike, was an interesting boss. I will always be in Mike's debt, though, for rushing me to Holy Family Hospital one morning around 5 a.m. when I was doubled over in the studio with what I thought was an appendicitis attack. Turned out it was stress-related. (Rewinding too many carts, maybe.) Nonetheless, the Walters family sent a get-well arrangement and let me take a few days off with pay. They were very gracious.

I left WYEN in September 1983 to go back to school full-time, and the guy I trained as the all-night replacement was Bob Worthington. A couple years later, when I was at US99, I talked to Ed about coming back and doing some programming and engineering work for him.

Sidney's conversation with Walters never panned out. Sometimes those conversations with Ed did pan out.

Cindy Linardos Bravos

Walters recognized a great voice from the very beginning of Cindy's career. Today, she has her company Bravo! Media, a commercial voiceover and narration company. Bravos writes for *The WYEN Experience* that she was working at WVFV in Dundee and set her sights on WYEN.

THE WYEN EXPERIENCE

This was my sophomore year in college. Louie Parrott at WYEN called. She wanted to see whether I was willing to do some fill-in work since the morning jock was on vacation. They moved the overnight announcer to morning drive and had me take the overnight shift. Mr. Walters stayed on my first overnight shift to make sure everything went well. WYEN was my radio home during Christmas and spring and summer breaks from school, and I did fill-in and production work for a number of years. I did middays from May 1985 to September 1986 when the station changed to the Z-Rock format, and I became promotions director. Those were great times in my life.

Dale Boe

Summertime 1986, Dale filled in for Cindy. Boe was hired off of his audition tape just to be used as a fill-in.

"This was unfortunately just as WYEN was ending and the format changing to Z-ROCK. At the time that WYEN hired me, I was working in Morris and was hoping my boss there wouldn't hear me on-air at WYEN."

Boe can think about this time at WYEN by looking at an old WYEN pay stub he kept. The talent fee was thirty-five dollars. Boe has worked on radio stations WCFL 104.7 FM in Morris, Illinois, and WDND in South Bend, Indiana.

"I was honored to work in radio for a dozen years and got to meet and work with some great folks."

Bruce Buckley

The night before Christmas Eve of 1983, the temperatures dipped to twenty-five to thirty below zero. Buckley parked his Chevy Chevette in the lot near the WYEN studios. He was inside the station working the evening shift.

"I put 'American Pie' on the turntable, ran out, started my car for a few minutes, turned it off, ran back inside, and played the next song. I knew exactly how much time I had."

Bruce didn't have another long Don McLean song, but he found what he could.

"The long songs came in very handy for bathroom breaks and warming up a nearly frozen car."

Until the shift ended, Bruce noticed the longer he waited between starts, the longer it took to start his engine, but at eighty below wind chill, he didn't expect the car to turn over without a fight. Would the gasoline run out or would his engine block freeze? Between songs, he had trouble concentrating on his show.

He zipped his coat and figured with the clothes he had on, he'd last a couple minutes outside, just enough time to put the key in the ignition and start the car he had turned on and off about a dozen times overnight.

He didn't need an instruction manual for starting a car in the cold. However, this wasn't cold; this was brutally cold—tundra cold. Bruce didn't own a manual on starting a car parked in an open lot all night in tundra cold.

Where one might hear an engine starting with a key in the ignition, Bruce heard nothing. He tried again. Nothing. Again. Nothing. He waited a minute. Nothing. Had this been a typical winter day, Bruce may have had time for thinking—but not in this dangerous weather. He only had time for reacting, and as cold air seeped into his car through the vents and frosted his windows, inside and out, he pulled his gloves on tighter, tugged at his hat, checked to make sure he'd zipped his coat. Bruce opened the car door, stepped out in the parking lot, and looked at the office door, more than twenty-feet away—and just then had this horrible thought.

What if the office door locked behind me?

He got closer ... saw frost on the edges of the door creeping up as though something with crystalline fingers and a palm had its hand on the glass and was working its way up the door. He finally reached the frosty handle, closed his eyes for a second, and pulled.

The door didn't open! He tried again with a harder tug, and it opened. His determination cracked the frozen seal.

"I slept in the lobby that night. Only things I wanted almost as much as a humming Chevette engine were a decent sized couch and better food than vending machine crackers."

Bruce called for a tow truck and waited and waited in the lobby, listening to pumped-in Muzak. No tow truck came. He waited. No tow truck.

Buckley's dad came for him in the morning. The tow truck never did. He and his dad left his car in the lot. A week later, the temperatures warmed, and Bruce got the car started and back home.

For Buckley, WYEN was an excellent training ground. On some stations, just reading the weather and briefly talking or doing sixty seconds of news wasn't enough. Bruce developed his radio personality by talking more than he would somewhere else. But in October 1983, he wasn't so sure he'd have a chance to work on WYEN. Bruce sent his tape and resume to then Program Director Mike Walters. Though Buckley didn't expect much to happen, he got a call three weeks later from Mike. Joe Cassidy had left the station, opening his position.

"I remember how nervous I was early on. The music was ending, and I didn't have anything to say next. My mind wasn't in sync with the demand of following songs with something to say, because I hadn't done this before. I think I was wooden on-air, and I believe Mike told me I sounded fine, but I think he knew I hadn't developed a personality. I can laugh now, but I was just doing the basics and not trying to screw up."

The big lake behind WYEN at the O'Hare Lake Office Plaza offered Bruce a little stress relief. Not only was the building more like a hotel with its crystal chandeliers and carpeting, but the lake looked very blue and comforting.

"One night I was feeding the ducks and geese with crackers. I put on a long song for a short walk from the hall. Security wasn't crazy about this because we attracted more birds and more of a mess afterward."

During the time Bruce worked at WYEN, the technology was changing fast. Stations were already bringing in computers, but Buckley described in 1983 to 1986 pretty much the same newsroom and on-air studio that were there from 1976 to 1979. He had the Selectric typewriter in the newsroom, and all the songs were on carts, and the turntables were still played for occasional album cuts.

"We still had the big, noisy, smelly, dirty teletype machine for the newsroom. Computers were just coming in, but I didn't see any evidence of it at WYEN."

WYEN was struggling financially toward the end, not unlike Bruce, making five dollars an hour, though he admits his salary represented a big raise from his four dollars an hour from Darrell Peters at WSEX in Arlington Heights.

"We had a few loyal retail spots on-air, but I could feel we were winding down, just had that feeling, and it turned out that way."

Bruce had so many memories of WYEN, thanks to Ed Walters. He praised Walters for being kind of a father figure and could easily see that Walters enjoyed the people working for him.

"I had fun talking to Mr. Walters about his experiences, and remember he'd say to me that sometimes in the radio business, you are not so much a broadcaster, but a circus carny. You're in the carnival business, he'd point out."

Bruce also got along well with the WYEN staff in the last few years of the station's run. He remembers some of the key people there: Kevin Jay, Louie Parrott, Jeff Dale, and Angie Smith. Each of them, he knows, got a shot at live radio because of "Walters's dream of starting up a station, and he did it."

With Columbia College Chicago under his belt and his WYEN experience, Bruce came to WLIT Radio as an on-air personality from 1988 to 2000, followed by WYLL/WZFS Radio, and eventually worked his way into a partnership as a graphic artist with Chicagoland Sign Company, a technical producer for WCEV Radio, and a part-time traffic reporter for WGN Radio.

Jeff Dale

Jeff Puffer is his real name. Today, he's a senior communications consultant for Frank N. Magid Associates, Inc. At his start in radio, Dale worked on-air at Request Radio in August of 1979. Jeff wrote his story.

The influence of FM radio was gaining enormous traction at that time. Long-standing champions WLS 890 and WCFL 1000 had set a standard for radio delivery styles in Chicago. The fast-talking, voice-bouncing (some would say "puking") announcing style that had been the norm for AM radio translated only in part to FM, the newer side of radio. In fact, a former WCFL disc jockey by the name of Paul Christie had the vision to recognize the changes in taste among listeners. He also saw an opportunity to consult radio staffs to help with the evolution to a more authentic, genuine, and conversational style of delivery. Paul made the decision after moving back to his hometown in suburban Detroit to hang out his shingle as a consultant. The heart of his strategy was basically to start at the top by establishing his trade and his reputation in the major markets. Doing that would ensure he would have a broad base of prospective clients in the remaining medium and large markets.

WYEN figured into his strategy because of our location in the number-three market. What's more, we weren't already committed to an agreement for market research or consultation with any other firm. As I understood the arrangement, Paul so eagerly wanted to establish a presence in the Chicago market that he was willing to provide the service if the station would cover his expenses. I am sure the station owners, Ed Walters and Jerry Westerfield, couldn't afford it. Keep in mind, Request Radio (WYEN) was not exactly a cash cow because the Arbitron ratings only flirted with "1's" out of a total, possible pie of one hundred pieces. That price—cover the expenses—was just right for WYEN. In the spring of 1981, Paul Christie began to sample the programming for music mix, rotation (i.e. how often various songs or categories of songs could be repeated), commercial break locations, and music "wheels" that adapted and adjusted programming tactics for various day parts of the broadcast schedule.

Another project Paul implemented was to coach, consult, and guide the air staff on the delivery style—how to speak and what types of "color" content to introduce on-the-air in ways that create and reflect a unique brand. Wayne Bryman, Nick Farella, Kenn Harris, Mike Roberts, Louie Parrot, and I all met, first as a group, and individually a couple of times with Paul. He knew what he wanted to hear; he had an instinctive, intuitive feeling for what he was aiming to develop. Paul even used his aircheck samplings to let us hear examples of where and when we were "doing it right" and where and when we missed the mark. In all fairness to Paul, he was never unkind or impolite when consulting us individually. He just couldn't effectively put into plain, simple English what he was hearing and thus wanted us to do. Only after having gone on for graduate studies in interpersonal communication and parlaying that into a specialty of my own did I finally recognize what Paul Christie was attempting to tell us. I've now been involved for twenty-seven years as a presentation talent coach serving television news clients worldwide, and I appreciate what was in Christie's head. I can say in Paul's defense that he understood the art of this kind of communication, but as art, it tends to be much more abstract—esoteric. It is not science and is harder to define or quantify. And to WYEN's credit, most of the people on the staff whose names I mentioned above did begin to develop an intuitive sense for the new form and the new sound of radio announcing—something different than the goofy, stylized-delivery of the earlier Top 40 era. This was one step we had taken that had some real merit in helping us to define a unique brand. One more "what if?" that pushed the envelope of WYEN—but evidently not far enough to put us in the running with the big-dog stations located closer to Michigan Avenue!

Bob Walker

"Mountain Man" used to produce little vignettes on the radio, and he did all the voices. Walker created stories and did all the voices, and then he'd play the tape and answer on the microphone just like all the participants were right there talking. He was one of the most creative announcers I've ever known.

Dan Diamond

Dan Lusk doesn't quite have the same sparkle as diamond, but it didn't much matter. The guy flat-out sparkled on-air. We talked a lot during the time we shared at WYEN. I always felt Dan was excellent on-the-air. His delivery was smooth, and his voice was easy on the ears. I wish I'd found him in all my searching, but I could not find Dan or his brother, Pat Lusk, a former part-time weekend announcer at WYEN.

Chris Devine

Devine is the essence of a true entrepreneur. Chuck Hillier worked in radio sales for WYEN and has known Chris. Chuck says he's been a sensational marathoner, a participant and owner of many of the events themselves. He's founder of Marathon Media and dozens of affiliated companies. Hillier says Devine has owned a great deal of radio stations in his time.

Reynolds and Meier mentioned that Chris is Joan Wrigley's son. His stepfather was Bill Wrigley Jr., a former owner of the Chicago Cubs and CEO of the Wrigley Company. However, Garry and Rob were clear in stressing that Devine made it completely on his own.

"He's definitely an overachiever from his WYEN days," complimented Rob.

Meier tells the story of how Chris once invited him to his home on the famous Astor Street in Chicago.

"I walk into their place, and there's Wrigley sitting behind this big desk. I went and shook his hand and thought I really made it now because here I am, some guy from the southwest suburbs, and I get to meet Bill Wrigley and go someplace on Astor Street, and for a kid that grew up on the South Side, this is pretty big stuff."

What might not be known about Chris is something he did almost every evening he came into the WYEN studio for his show. As I was finishing my evening newscasts, Chris had this ritual that struck me as generous. Every time he'd see me, he'd have a cup of 7-Up for me. He bought soft drinks in a vending machine down the hall, and the drink came in cups. I guess I must have looked really thirsty. I never questioned his generosity.

Chris did not wear socks. He was the sockless man. He reminded me of James Dean, the legendary actor.

How about the evening announcer without the socks? I asked Wayne for his take on Chris.

"Interesting guy, very entrepreneurial, and you are right, he never wore socks." Wayne thought a bit more about Devine's success in the radio business, noting that he had a lot of respect for him.

Devine was the coolest guy at WYEN.

CHAPTER TWENTY-FOUR

KEEPING THE DREAM ALIVE

In the 1970s, many small- to medium-market radio stations were still owned and operated by individual radio station operators, before large corporations swooped in and carried away station after station, feeding listeners a new radio landscape. Ed Walters was by no means unusual this way. He was not part of a corporation but a hands-on owner working every day and interacting directly with the staff.

In my career, I experienced a variety of station ownership ranging from a family whose patriarch was local legendary broadcaster Mal Bellairs to a station in a house at the end of a block whose owner was once Chicago Cubs broadcaster Jack Brickhouse.

During Walters's years as an owner, the on-air talent and sales staff saw the boss all the time. We knew how the dollars were stretched and how much money in spots the sales staff had to sell for our station's existence.

Every radio station and business faced the same "nut to crack," as Walters was known to say. Dozens of sales people came through WYEN for nearly fifteen years. One of those sales people was **Chuck Hillier**. He tells his own story.

Hey, gang, wow, what memories! Walters told me WYEN stood for What Your Ears Need. WYEN was my first sales job in radio before I jumped to the national rep side of things and ultimately to WKQX-FM, Q101.

Yeah, the pay plan was pretty dreary: straight commission, seven percent, payable on collection, no base, no draw, and no expenses. Top spot price at the time was twelve dollars. I drove up and back every day from Beverly on the South Side. Whew, thank God for cheap gas back then.

I was in contact with Meier, Reynolds, Devine, and Brown. Greg worked with me at Q 101. That is indeed Rob's voice behind the Speedway spots. Reynolds is also the voice behind dozens of auto dealers in the area. He and I partnered in New Standard Communications, the developer of the "Smokin' Pop Standards" format, which played as Red 104.1 in St. Louis as well as Jeff Trumper's Star 97.5 in Phoenix. Rob, Garry, Chris, and I would schedule lunches to get caught up a couple times a year. Ron Leppig was general sales manager, and Bob Sparr was on the sales staff. Sparr joined the sales teams at a couple of downtown AM stations before moving out of the biz. Bonnie Baker also learned the sales craft there before getting into the rep arena as well. I remember when Arbitron told Walters that WYEN actually qualified for the book. He bought a couple-report package ... and there we were, with something like a .2 share! I punched out sales pieces proclaiming that we were "The Fastest Growing Radio Station in Chicago!" (Well, from "nothing.") Further, because we were listed last (alphabetically), we used to turn the book upside down and claim that "We're number one." Seriously, though, however teensy the audience was, I can tell you the TSL (Time Spent Listening) was astonishing! Never out of the top three, sometimes number one, that's remarkable.

Hillier is high energy and very persuasive. If there had been a demand for a model radio sales executive, I'd have asked Chuck to step forward for the role. The next account executive in *The WYEN Experience* agreed with Hillier on the commission structure.

Jack Johnson worked at WYEN on the sales side from 1977–79 and also did a little bit of production with his voice on a few spots. Johnson says WYEN Radio stood for *W*here *Y*ou're *E*verybody's *N*eighbor. In his story, Johnson recalls his WYEN experience.

WYEN was my last radio gig and the one that convinced me to get out of the business and into the parallel universe that I've been in for the past thirty-one years. Walters also wasn't a bad guy by any means. Same goes for Westerfield and Mrs. Walters. Despite having a few Walters family members on the payroll, the place didn't have the classic nepotistic feel.

Life in sales was something of a grind.

The two people whom I reported to were knowledgeable pros, but not the easiest to work for. I'll just leave it at that. They also had one of the worst commission structures I've ever had to work under.

One thing, I will say about the station—it sounded good!

THE WYEN EXPERIENCE

Johnson recognized a very tight, very professional sound from the announcers and a good jingle package.

Johnson and **Pete Hoffman** both worked under Sales Manager Ron Leppig. Hoffman told me of how WYEN was his first job out of high school and how early in his sales career he'd sell air space to an airline and to a boxing promoter.

Hoffman could answer this question easily. Does the fight game and sleaze go together? How about the fight game, sleaze, and two thousand dollars for advertising a boxing promoter's matches? Hoffman met a boxing promoter offering to pay a couple grand for advertising time on the suburban radio station.

"We met for lunch at a restaurant in Des Plaines on Higgins and River Road, where flooding was a recurring problem. I remember because a public fountain marks the spot today."

Long before a crew tore the building down because of water damage, Pete and the promoter sat and talked about *the show*, and the guy handed him a wad of cash.

"I thought *holy crap*, but I tried staying cool, like this sort of stuff happened to me every day. I doubt I pulled that off, but apparently I pulled it off well enough that he didn't seem overly concerned about handing me two thousand dollars."

Pete knew this wasn't a trivial buy on the radio station! He brought the money back to Carol for advertising this battle of the heavyweights—a dozen boxing matches at a Rosemont hotel.

"I walked into the WYEN office, and Mrs. Walters was there and saw me. She said, 'It's all kind of *sleazy*.' I concurred, but I shrugged it off and pulled out the two thousand dollars."

Except for music filtering through the closed studio door, the chatter from the office staff dropped off, letting the word sleazy disappear.

An ad buy from Lufthansa startled the staff, but Pete knew from some industry publication that the airline was intent on buying spot time in the Chicago market, so he called the media buyer in New York and said WYEN covered a wide suburban area around O'Hare Airport. Hoffman knew of course that people who travel a lot buy homes within easy reach of an airport. Though Pete did his homework on Lufthansa and flying habits, he felt the media buyer believed he was some "knucklehead kid on the phone." However, she also recognized the amount of money he asked for did not exactly create a huge hole in her budget.

"I remember our sales manager was quite surprised I swung the deal."

Hoffman worked for a brief time at WYEN from 1978 to 1979. Pete talked to me from Toronto, where he called on a number of interactive ad agencies in his job as director of sales central for Bunch Ball.

"I had worked briefly at one of those insidious telemarketing firms that sell radio safety messages basically out of a bullpen (phone bank). Did that for three months, and I frankly learned how to get people on the phone."

Then finally WYEN came about. Right after college at the University of Missouri, where Pete majored in journalism, he learned how to make contacts and had an interview with WFYR-Chicago, but this didn't pan out. A high school friend provided the key to his future, or more precisely, the key was his high school friend's father. He'd been a marketing vice president in the international division of McDonalds. Through his connections, Pete talked to agency people, who in turn introduced Pete to Leppig.

"I was a finalist the first time around interviewing with the sales manager, and after six months in which I talked quite a few times to Ron, I got hired. I did not make a lot of money at WYEN. I found WYEN typical of the radio industry in general. WYEN did not have an organized training structure. You were given a rate card and told to go sell, but Leppig came along and helped guide me through the sales procedures."

WYEN sent Pete to sales seminars. However, in terms of structured training and teaching, he conceded it wasn't much.

Although Hoffman wound up for a time as top sales person at WYEN, arriving there was not easy. Early on, he hadn't hit his sales numbers, leading Ron to question whether Pete could remain on his sales staff. However, Ron saw something in Pete and convinced Walters to let him work it out.

"They kept me on, and I wound up doing quite well for WYEN. If Ed was truly looking to fire me, he never made me feel like I was on a clock or anything else like that, so I have warm feelings for Ed and Ron. I thought the world of them."

The sales office was down the hall from the radio station offices. In my interview with Pete, he told me his spatial memory is pretty good, so I asked him to recall the sales office and where everyone sat in 1978–79.

"Leppig had the office, and the rest of us were in a bullpen. I sat right outside Ron's door, Ed Peters sat next to me, Bruce Krawetz sat next to him, and Dick Runtz sat in front of Bruce, Greg French, Jack Johnson, and Bob Cronin."

Pete threw in the fact Cronin's twin brother was the father of REO Speedwagon lead singer, Kevin Cronin.

"This was a terrific first job at a small family-run company, but it wasn't the sort of place you could build a career. However, WYEN was a fun environment in which to work and learn the trade and make a couple of bucks. Working at WYEN helped me secure my next job, and pretty much every one of my sales coworkers gained from their WYEN experience by eventually moving on to a bigger position."

That boxing promoter with the mod, grey three-piece suit and flared pants with the wad of cash had quite a boxing card as Pete remembers more than thirty years ago.

"I went to the event because I'm a big boxing and martial arts fan. Sylvester Stallone's *Rocky* was still an influence, and there was this local Italian heavyweight on the main card. He won his fight, and we heard the *Rocky* theme on the public address system all night. Six months later, on Rush Street, I saw the same promoter with a boxing card at Mother's. However, after the Battle of the Heavyweights in Rosemont, he was done with the big events and advertising on WYEN."

The radio station gave Hoffman and **Sherri Berger** wonderful opportunities. While Pete learned the business of sales and set his future course, Sherri added voiceover work to her fine selling skills.

As I prepared for my interview, I remembered three distinct things about Berger's time at WYEN. Sherri was one of just a few females in sales, voiced lots of spots (one with the F bomb), and hosted a holiday party where I got to bring an up and coming comedian.

Berger had reason for her uneasiness the day a coworker called, telling her he had just heard a commercial on WYEN and could swear it was her voice saying the "F" word. Was the coworker kidding? You read about these things happening in the business, but not to professionals of Sherri's talent in voiceover work.

On the day of the *F bomb*, Sherri sat in the production studio voicing a spot with help from a producer, but all these years later, she can't recall which staff member was directing her performance. She remembers making a couple of mistakes in her reading and getting so frustrated and then uttering the F word. Normally anything extra like a swear word is edited out by the person doing the editing.

"The spot aired with the F word in it!" Sherri was stunned, scared, and wished she'd used the word *crap* or *shoot* or *dang it*. Her phone rang again. Instinct kicked in. She knew Mr. Walters was calling, and she had this overwhelming sense of responsibility. She remembers Walters absolving her and deferring blame on the producer.

"The editor was in deep trouble," Sherri cautioned. "I knew the FCC could penalize the station because their investigators, I believe, bore down much harder back in those days than they do now. Not that the FCC would threaten me, but I felt so responsible in the way the person handling the production work might get fired or the station fined over it."

The sales manager waited for the first sales meeting after the incident and gathered his sales team, describing to them the Berger gaffe. The F bomb incident burned right through the station, some knowing immediately that Sherri's voice was on the spot; others never knew the whole story.

Sherri survived the F bomb incident and went on to success as a WYEN account executive.

"I promised my clients the world and delivered it too. I wrote their copy, guided them with the times for their spots, and gave them the best deal I could. The staff under the direction of Ron Leppig gave the clients a lot of service and more attention than they'd get downtown."

For WYEN's sixth birthday, Sherri invited the staff to a party at her apartment in honor of the station anniversary and the holiday season. I brought *Men are Slaves* comedian Judy Tenuda, without her accordion, to meet Meier, Reynolds, and other WYEN staff. During my interview with Sherri, she asked how I knew both the date of the party, December 15, 1977, and the location of the party, International Village Apartments of Schaumburg. I told her I kept the inter-office correspondence from December 6, 1977. She thought she was bad, but now she knew someone who kept even more junk. Yet Sherri was one of the few people I can recall from my WYEN years that found ways to engage the staff in events outside the station. Of course Ed was the other with his softball games.

Between WYEN and carving out a wonderful career for herself in the voiceover industry, Sherri worked for HR Stone, which led to starting her own ad agency with clients she worked for at WYEN, including Erewhon Mountain Supply. She went on to pursue film repping and voiceover full-time. The road to owning Voice Over U started at the Learning Annex, but after it closed suddenly, she started her own training program on a very small scale. An opportunity to teach classes through Act One, a prominent Chicago acting school, offered a chance for growth. With the advantage of becoming very well known in the voiceover industry, and gaining an excellent reputation as a teacher and coach, Sherri was ready to expand her training program in the early 1990s and officially opened Voice Over U www.voiceoveru.com

People interested in voice work as a full-time or part-time career have come through her doors in Chicago. She's helped so many talented people navigate their way into the voiceover industry. Ask her about student success stories, and she'll proudly state there are many, including a top promo announcer who earns six figures in Los Angeles and many others around the country getting work in all genres of the business.

"Many out-earn me! Past students keep in touch with calls, e-mails, and letters attributing their success to what they learned studying with me; that is my reward. My opportunity at WYEN and encouragement from Ed gave me the confidence and experience to create a career in voice-overs."

THE WYEN EXPERIENCE

The Berger gaffe is a learning experience.

For broadcasters, fear of swearing on-air is as real today as the fear was in the 1970s. However, stations are generally regulating themselves, warning their employees that profanity is not acceptable, will likely cost their jobs, and can be expensive to the individual and station. A complaint generated by the public is the usual method for an FCC investigation of potential profanity, obscenity, and indecency. Still, broadcasters remain aware of the consequences, and the majority choose to steer clear of anything that'll prompt FCC attention to their station.

Because this whole subject is not taken lightly by broadcasters, they make sure they don't slip on-air. Some have special techniques while others just wing it and hope they don't say something they can't take back. I may be in the minority here, but I've gone the extra step, and although this may sound somewhat crazy, I've developed a list of *alternative words*; a substitute for swear words. I began some time back substituting those words into my everyday vocabulary. Where other broadcasters keep the swearing internalized, they won't have any use for a list. I've noted the longer I stay in broadcasting, the longer the list grows.

I developed the use of *dang it* in my limited anti–swear word vocabulary, but *dang it* didn't bear true from my mouth.

I heard my share of *shoot*, but I personally used this *alternative word* sparingly, choosing a dramatically emphasized *oh crap* for a myriad of things gone wrong at WYEN. Jocks would mistakenly start out saying *shi*, but they'd quickly slide in *oot*. This sounds as though the announcer is saying shoot. *Shioot* is a fabricated word meshing a swear word and a regular word, creating a use for a non-existent word I've termed a *slide word*. It slides right into a sentence, though the discerning can tell your intention was likely a four-letter word.

FCC penalties can be substantial. The threat of a fine from the Commission and potential wrath of the general manager are effective deterrents today just as the WYEN on-air staff faced in the 1970s and 1980s. Losing your job was a real possibility on WYEN, but that was true throughout the business for saying a word on-air you may have said many times outside the radio station without fear of reprisal.

Dick Runtz never worried about saying words frowned on by the FCC. His focus has remained sales. He'd learn how to turn a *no* into a *yes* from a client. Dick hadn't yet learned the techniques for persuasion or collection the day he walked into WYEN and talked to Walters and Leppig. However, the Southern Illinois University graduate had tried to gain experience in sales.

Before Runtz left his job at WBBM in traffic and continuity, he approached the sales manager of WBBM AM & FM.

"The sales manager offered encouragement," Dick recalled, quoting his specific advice, "Do well at a smaller station. Then come back and we'll talk."

This didn't terribly surprise him.

The sales staff at WBBM might have advice Dick could use, and he surely wasn't shy asking. One of the guys suggested WYEN and mentioned Leppig. He contacted Leppig, and they met at the station with Walters. He felt the meeting went well. Dick got hired for sales in 1979 and stayed fifteen months.

This was the disco era, and Dick remembered selling to a number of retail clients that catered to disco.

"I remember selling to a store at the Tin Cup Shopping Center at Golf and Algonquin Roads. There was this little store called Pants and Dance Disco Wear. There was another place called the Navarone Disco on Higgins Road and Interstate-90, and I got them to advertise."

The key to the smaller merchants, Dick recognized, was in working with them in producing their on-air spot. This is how he learned to sell.

"Clients not interested in advertising at first might rather have an advertisement in the newspaper they could see and feel. However, what was really kind of funny was that I'd return to their store with a cassette recorder, and I said we'd put together a sixty-second spot, no obligations. I gathered information on what kind of merchandise the client would like to move and put on special, and then I returned to WYEN and had the spot produced, narrated with a little music bed. I returned with my little cassette recorder and played the spot for the client. The first reaction was that they no longer said they didn't have time for me."

Here's one conversation that followed hearing a spot:

"You've got the address wrong."

"Okay, we'll fix that."

"Well, I don't really like the music."

"Well, what kind of music would you like to hear in the background?"

"Something with a disco beat."

"We can do that."

"All of a sudden," Dick emphasized, "they are involved in producing a commercial, and you make the changes and come back with it, and now they want to hear it on-the-air."

Now the client is in show business.

One of the downsides of sales, whether radio air time is sold or some other service, is that sales people hope the client pays for the merchandise on time. Otherwise, in many instances, they won't receive a dime of commission. They

THE WYEN EXPERIENCE

also have to go and collect money if the client is reluctant before they turn it over to a collection agency or as a last resort go to court.

Dick had never collected. All that was about to change the day Walters called the sales staff into a meeting. He had a whole stack of paper with dozens of client names and amounts owed. As Runtz recalls, Walters put the stack down and turned to his sales staff.

"If you guys want to follow up on these, I'll give you 10 percent, whatever you can collect."

"I'll take a handful," Dick remembers saying to Ed.

Dick headed to Woodfield Shopping Center in Schaumburg for a men's clothing store, where the owner had not paid his bill to the station.

"The owner tried to ignore me. I spent half of the day loitering in his shop. I'm talking fairly loud about his bill and that he's not paid. His customers can hear every word. He finally wrote me a check so I'd leave. That was my first experience going out and collecting."

Not getting stuck inside the office really appealed to Runtz. He spent some time in the building, but was encouraged to make as many personal visits to potential clients as possible, and he enjoyed every moment. However, the down side became too much to overcome.

"I left on my own because I wasn't making any money. This was disappointing, and I thought maybe the radio sales business wasn't for me."

Today, Dick works for a biotech company, but he says, "Never in a million years when I was in college would I have guessed that I'd be doing what I am doing now."

His road after WYEN took him to a twelve-month job as a purchaser, but he "felt like a caged animal," and then he began working for a telephone and data systems company that set up voice-paging service. He talked to physicians, selling them pagers and beepers and voice-paging communications, and in the process, he learned about pharmaceutical sales.

"I ended up going to work for a pharmaceutical company around 1983, and everything kind of clicked, and I've been in the pharmaceutical industry ever since."

The best medicine for Runtz is the work he does with pharmaceuticals. For **Kathleen Cahill**, the best medicine for her remains radio sales. Walters saw to it that Cahill could survive anything.

"He discouraged me as much as he could by bringing out the negatives," recalls Cahill on her interviews with Walters for the job of WYEN account executive. Walters wanted to make absolutely sure she understood what her first sales job in radio would cost her emotionally and the kind of stress she'd encounter. The job interviews were spread over three weeks in 1985 with

Walters. They met six times. Cahill confidently pushed forward, promising Walters she'd bring in a car dealer she knew well.

"The promise sealed the deal, except Walters stayed on me until I did bring in the car dealer. A month later, Walters said there were too many spots on-air from the car dealer, and he was tired of hearing them. He now wanted me to dump the order."

One day Walters had big news and asked for Kathleen's help in gathering the staff for a round of drinks at a place called the Shady Rest.

"Once we were ready, Walt directed his attention to us, looking serious, telling us we were replacing our music with heavy metal from the Satellite Music Network."

A normally enthusiastic group put down their drinks, stared straight ahead at Walters, trying to catch something in his face belying what he'd said. Nothing appeared—not a wink or head shake, not even a wry smile.

"I thought he had lost his mind since the format seemed so far from his style, but he was over the top with enthusiasm about it."

Cahill came so close to parting company, but something changed her mind. Maybe her intuitive skills sparked her new view on heavy metal, thinking maybe this was a good opportunity.

"I scarfed up all the male-skewing accounts I could find. We were booking orders, making a lot of money. I had Budweiser on and was prospecting accounts that made sense. We went on for quite a while with the heavy metal format."

During her time at WYEN and eventually Z-Rock, WZRC (FM 106.7) Des Plaines, Cahill had several contacts from WLS and other downtown stations. But the convenience of geography influenced her decision. She lived a couple blocks away from WYEN and wasn't moving on … just yet. A couple of events helped make her a little more willing to change her mind. The Great Flood and another format change outweighed distance considerations, and Cahill finally picked up on WLS.

The Great Flood started simply as steady rain one day in 1987. The staff noticed the retention pond at the O'Hare Office Plaza couldn't hold the rain. But without any sandbags around the pond, the level rose, and the water eventually rushed across the grass and poured into the buildings surrounding the pond, introducing unwanted visitors with gills and fins into a lower-level office where WYEN was located. The WYEN staff felt fairly helpless as the water soaked both the carpet and the walls, seeping under the doors and through the cracks in each of the offices. Kathleen couldn't tell whether the water from the retention pond might completely submerge the on-air studios and sales office, but the water level had reached three feet, and the rain hadn't stopped. Cahill says she found no evidence of sandbags.

"Carol Walters went back and forth between the radio station offices and a lower terrace looking out at the pond. She and Walt were vulnerable because they still were involved in the station after selling WYEN to Vern Merritt."

Kathleen and a couple of coworkers left for lunch and came back no more than an hour later.

"I couldn't walk down the stairs to the radio station because the water was already several feet deep, and the hallways' emergency lights were on. Besides the fish swimming through the hallways, several people splashed through the filthy water in wading boots, but most of us weren't equipped. We just couldn't wade into the water. Everything appeared submerged."

The next day, the water level dropped several feet. Kathleen saw thoroughly soaked hallway carpets and water stains on the hallway walls, and she squished her way to the offices. Pretty much everyone worked on saving equipment and files, salvaging what they could.

"I didn't have hip boots, but we were asked to put on a pair of gloves because of the toxic mess. I helped remove my files from the sales office and put everything on the second floor where we'd set up an office. Oh, I also had to go in for a tetanus shot at Lutheran General Hospital in Park Ridge."

Once Merritt had control of the station, he brought in Craig Wilbraham as his station manager.

"Vern had religious misgivings," Cahill recalls. "I heard he was also consulting astrologers. They told him he should quickly drop the heavy metal format. He did within days and now had The Wave, WTWV. Problem was that WNUA, a much bigger station, had started a similar format a month earlier. Really didn't make any sense to me to take up the wave format (New Age)."

The Great Flood, and two format changes later, the inquiries made from WLS were looking better and better to Cahill. She finally made the switch.

Cahill eventually went to work at The Lite 93.9FM in Chicago and "occupied every seat in the house from sales to general manager." After eleven years at The Lite, Cahill started her own business as a custom home builder and now works as a radio sales manager of WZSR-FM and WWYW-FM.

Chapter Twenty-Five

THERE AT THE END

Those Greg Stephens parties in the late 1970s and early 1980s in Carpentersville netted a long friendship with Terry Flynn. Not only did Terry and I have broadcasting in common, but we also were friends with Greg and all the other people Stephens was close to in those days.

Terry worked at WYEN in 1985, nearly the end of Request Radio. This is Terry's story.

I got my first tour of the WYEN studios at the invitation of Jack Stockton, my mentor at WWMM in Arlington Heights. Jack had just left WWMM for the evening air-shift at WYEN. I found the station to be very functional, although nothing elaborate as far as studio design or equipment. Still, they were getting the job done all under the glow of bright fluorescent lighting. This was a full-service, adult contemporary music station that didn't shy away from informing listeners with hourly newscasts, sports scores, and drive-time traffic reports. Ed Walters allowed disc jockeys to have a real and lively on-air presence.

Flip the calendar forward, and the year was 1985 when I was wondering if there may be anything available for me at WYEN. I made a call to Larry Northon, whom I knew from college and who was working there as the station's news director. Larry put me in touch with Walters's son, Mike, who was then in the position of program director and handling the hiring of air staff. That led to me being hired for some vacation fill-in, followed by the 7 p.m. to midnight slot, six nights a week. In addition, there were commercial production duties each night

THE WYEN EXPERIENCE

that I would do for a couple of hours on either side of my air shift, sometimes both sides depending on how much was sold.

ACTOR AND FILM DIRECTOR RON HOWARD TALKS TO WYEN ANNOUNCER MARK DIXON.

I never actually met Mr. Walters, whom we all knew as Walt, until I was fully employed at WYEN. While I was on-air, he would often be around the station into the early evening. Walt would occasionally enter the studio to check some paperwork pertaining to the copy book, transmitter operating clipboard, or a recorded commercial in the cart rack. Those visits were always with a purpose, to get in and get out, and I never felt any extra pressure while he was in studio. The vibe I got was that he trusted me to execute the sound of the station the way ownership and audience expected it, and he got out of the way to let it happen. If I recall correctly, I was working at WYEN for a few weeks before starting to have conversations with Walt about radio. My first impressions were that we would never have these conversations. I then found Walt to be a very accessible general manager and a very down-to-earth person. I loved talking shop with him whenever there was an opportunity to do so. I wasn't sure where the station's level of success was at the time I was hired, but Walt indicated to me that even though business was down, he was optimistic good days were still ahead. I'll never forget how he expressed that. He told how he wanted to "give it one final push," and if those aren't his exact words, they're close. With that phrase, he made a slow fist pump motion forward with his arm across his torso, not into the air above his head, but forward and across his body. I was encouraged by that and was glad to be working with him toward achieving that goal. Sadly, the Request Radio format would change to a couple of emerging, satellite-delivered formats of heavy metal rock named, "Z-Rock," and new age jazz called, "The Wave."

The one conversation I cannot remember ever having with Walt would be the one where he would tell me how it all began. As I was breaking into the business at WWMM, WYEN had already established itself as a full-service and fully-staffed adult contemporary music station. The radio station's signal covered the north, northwest, and west suburbs very well. WYEN was the station of choice to be played in so many shops, styling salons, and home-entertainment departments on demonstrator stereos. I wondered whose idea it was to make WYEN sound the way it did when they first hit the air. WYEN benefited from hourly newscasts and traffic reports from a full-time news department. The station also had tight program elements, station jingles that were long and lyrical, friendly sounding personalities who were allowed to speak often and over the intro's and outro's of the songs they were playing, and a well-known phone number that a listener could use to make a song request or win a contest.

I did know that Ray Smithers was the first program director of WYEN. Smithers also had something to do with WWMM's predecessor, WEXI.

In the late 1960s and early 1970s, WEXI was known primarily in the northwest suburbs as the station that was playing all the hits that could be heard on WLS and WCFL—but with the advantage of FM stereo quality. Even though no live disc jockeys were on-the-air, the station had a lively sound that was not typical of FM radio in those days. Similar to WYEN, the program elements were tight. WEXI's own announcers identified song titles and artists, delivered regular newscasts and weather reports, and voiced commercials. Was it a coincidence these two stations introduced a brighter, more contemporary sound to the FM dial, and Smithers had something to do with both of them consecutively?

Random thoughts:

While I was on staff at WWMM, I recall thinking how ironic it was that we could gaze out a back window of our building, and there was the WYEN tower seemingly gazing back at us. WWMM had a beautiful broadcast facility that was state-of-the-art for its day. WYEN had the powerful signal we wish we had. It made for a David and Goliath type of competition, although I learned later that we at WWMM were far more aware of them than they were of us.

Some of the men who were working as announcers, automation and transmitter operators were kind enough to let me occasionally hang out at WEXI for a while as this aspiring broadcaster absorbed whatever I could about creating the sound that came out of my radio at home. At least two of them told me about a new local radio station that would be going on the air soon from Des Plaines. This was the first time I heard about WYEN. These announcers were excited about

THE WYEN EXPERIENCE

the possibility of working there because, as one of them said, "Live radio, man." Apparently they sought to branch out from the confines of the WEXI automated format, hoping to become the future live personalities on a station that did not even exist yet. WYEN was still several months away from signing on the air

I came to WYEN after breaking into the business and staying at WWMM, Arlington Heights for six years. While there, I was able to do part-time at Bonneville's former WCLR and WVFV, Dundee. That led to WAIT, Chicago and WXLC, Waukegan and WYEN in the mideighties. I did not plan it this way, but Arlington Heights was under different ownership and management. Worked well for me once and worked well again. Then two years at Chicago's FM 100, and throughout the 1990s, I worked concurrently as a traffic reporter on WBBM Newsradio 780 and as a staff announcer for WTTW-TV. I fill in on news for WDKB, DeKalb.

Adrian Sakowicz met Mr. Walters in February, 1981. Sakowicz tells the story of how he first sent a letter and a tape pitching Walt on joining WYEN in a full-time sports reporting role.

This would be a first for the station as a full-time sports reporter and anchor. Walt's son, Mike, joined us for the meeting in the WYEN sales office down the hall from the studios on the lower level of the building. Walt listened to my tape and then started writing furiously on a piece of paper, occasionally using a calculator during the process. After a while, he stood and said that even though the station needed to concentrate on music, he thought his staff could sell the hourly sports shows and create another revenue stream. Thus, Sakowicz on Sports was launched and began airing a couple of weeks later with three hourly reports in morning drive, and a month or so later adding three reports in afternoon drive. It might have been my sports knowledge or my broadcast training at Marquette University, but I also think Walt may have just been impressed I could spell and pronounce his real name—Piszczek.

While the station's focus was clearly on music, I think my timing was perfect. WYEN began airing the Chicago Blackhawks games earlier that season, and Walt was looking to strengthen the relationship between the broadcasts and the drive time audiences by airing game highlights during my sportscasts. In an era where many of the sports teams were in need of broadcast outlets, Walt had figured out that one way to broker large chunks of air time was to offer it to local sports teams so they could guarantee a season-long play-by-play presence of their games. Paying a team big dollars for their broadcast rights was reserved for only the chosen few. Pat Foley did the play-by-play for the Blackhawks in what may have been his first exposure in the Chicago market.

I remember Walt calling me in his office one day and showing me a huge box full of cassette tapes from aspiring hockey announcers who wanted the chance to do the Hawks games. He played Foley's tape from what I think was a minor league game, and I remember us both agreeing that this guy was going to be a star some day. How prophetic. Right around that time WYEN also aired Chicago Bulls games for a season or two, as I remember doing the pregame and half-time shows from the studio on several occasions. While the evening deejays weren't too thrilled to lose several hours of their shift to a sport's broadcast, it was certainly one way to appeal to another section of the listening audience. And as the sports guy from the Bulls and Blackhawks flagship radio station, it certainly got me a better seat in the press box on game days.

I vividly remember one of the true highlights of the era was the day the painters completed the huge WYEN Billboard on the outbound Kennedy Expressway around Irving Park. Walt brought in a picture of the sign, and it felt back then like WYEN was on par with any Chicago radio station, now that it had made a solid statement of authenticity on a major thoroughfare.

The WYEN on-air talent line-up in the early 1980s was extremely strong and truly a breeding ground for so many talented announcers. When I arrived, Jeff Dale (Puffer) was doing the mornings, Louie Parrott was on middays, and Nick Farella handled the afternoons. Wally Gullick did the news, and later in the 1980s was followed by Mark Hilan (later worked for NPR in New York), Larry Northon, and Kris Torres. On-air people like Bruce Buckley, Bob Worthington, Terry Flynn, and Phil Raymond also spent time honing their craft at the Des Plaines outlet Y 107, as it was called. The overall sound was clearly on par with anything else you would find on the dial.

Sometime in the early to mid-1980s, Kevin Jay came on board and handled morning drive. Kevin was very talented, and we both had a great time bantering back and forth in and out of the sportscasts. Many times my two-to-three-minute segment would turn into eight to ten minutes, as we easily got off track and started on whatever was the topic of the day. Invariably, Mike would be waiting outside the studio door to scold us that the station is about music and not cracking jokes. He was probably right, but the reprimand did not seem to bother Kevin and me. Walt had a great sense of humor, and even after Mike's lecture, he would look at me and laugh as I would walk by.

In my sports role, I was responsible for selling my own shows but often got help from the WYEN sales staff with whom I had to work closely. It was a sales staff that had many future stars as well. Ed Peters ran the group when I arrived, and I remember other pros like Bob Faust and Pat Kelly among the several young sales professionals later coming on board. All of these folks went on to big radio and television sales careers in Chicago and beyond.

THE WYEN EXPERIENCE

During the Major League Baseball strike of 1981, Ed Walters brought his softball team to Wrigley Field and played the Wrigleyville branch of the Chicago Fire Department. Among the players for WYEN, announcer Louie Parrott leans on a baseball bat. To her left is Diane Finkler. Behind Diane is Wally Gullick. In the back row to the far left is announcer Mark Dixon. On the right in the front row kneeling with a baseball glove is Walters' son, Mike. Behind him and to his right is Adrian Sakowicz. The team photo is courtesy of Adrian Sakowicz.

I was able to continue my Sakowicz on Sports role until sometime in 1989. For the last six years, I arrived at the station about 5 a.m. and reported a couple of sportscasts live before taping the rest of the day's feature items and timeless stories, leaving the actual score updates for the jocks to wrap around my sportscasts. With sports in Chicago getting a boost from the success of the Chicago Bears in the mid-1980s, this schedule allowed the station to keep a sports presence while enabling me to run off to another full-time job. I wasn't able to work with the crew at the station as much as I would have liked, but we were still able to meet up regularly at station promotional events and softball games.

One of the better known radio sports figures in Chicago during that time was Les Grobstein from WLS. Les formed a softball and basketball team called CRAS (Chicago Radio All Stars) which played in charity and exhibition games throughout Chicagoland. Virtually all of the radio sports guys played on it, including Dave Eanet, Pat Benkowski, Tom Greene, Walter Burch, David Schuster, George Ofman, Steve Olken, Jerry Kuc, Fred Huebner, and me. Les

STEW COHEN

was the player-coach and scheduled dozens of games. It was always very telling as to the power of the WYEN signal and listenership that the cheers for WYEN during the introduction were as loud as those of all the other big-hitter outlets.

Jay and I were among the longest tenured people to stick around through the end of WYEN and right into the 1986 transition to Z-Rock and not too long thereafter to The Wave. I still see Kevin on television these days selling insurance and other things. Even to this day, on occasion, I'll be leaving my name for a phone message or restaurant reservation, and just after I say the name Sakowicz, the person on the other end will say, On Sports. It's funny how WYEN lives on in small ways.

FROM THE WALTERS-HEINLEIN COLLECTION

After graduating from Marquette University with a degree in broadcast communications, I began working at WWMM in Arlington Heights for four months as the overnight announcer and four months as morning drive deejay. From March 1981 to June 1989, I worked first as sports anchor at WYEN, then WZRC (Z-Rock), and The Wave. I've also worked as TV play-by-play and sports reporter for WMVS/WMVT in Milwaukee and at WBBM-AM Newsradio 780 as a producer and writer. Also I've worked in Corporate Communications with two Chicago area Fortune 500 companies.

CONCLUSION: WYEN MEMORIES

The program director recruiting for a job or two for his radio station had only to find me in a room of graduating broadcast communications students in a Holiday Inn in Carbondale. *Big Breakitis* didn't infect only me. Each of us in the room of new student members of the Illinois News Broadcasters Association believed someone in radio had magically recognized our good qualities and was there to interview us and possibly hire us upon graduation from SIU. I was one of those students at the INBA Convention of 1976. We were attending seminars on radio writing, interviewing, and other areas of broadcasting that extended our radio education, but the number-one lesson learned was thrust upon us at the end of the convention. Each of us in our own way realized a professional career must start somewhere else—maybe Iowa or Nebraska or Oklahoma. While the INBA weekend was a very positive event for young broadcasters, few if any would land a job right there. More determination and doggedness of purpose were required. The convention was beneficial, nonetheless. Months before I started in the WYEN news department, a candidate for Illinois governor convinced me to stick with broadcasting. Jim Thompson grasped our hands as he worked our room at the convention. Thompson didn't know the effect he was having wasn't so much about his good qualities, but about something else entirely. As the Republican candidate walked slowly down an aisle shaking hands, it became apparent our very presence as budding broadcasters generated this type of interest from others to seek us out. We wanted more.

Big Jim exuded energy in movement and clearly enjoyed talking to us. We twisted from our hips, looking up at the tall US Attorney for the Northern District of Illinois. His sense of right and wrong was evident. His eyes showed a hint of the steel will in his heart. He drove forward, touching a few shoulders along the way. He had prosecuted crooked politicians, and we appreciated his work because his efforts made for great news stories we read at the city, college, and campus radio stations where we trained for *the big time*. He wanted us to meet him and like him, and if we ended up working in professional broadcasting in Illinois, we'd show him a little positive attention. I'm sure he couldn't see the future, but Thompson was fairly intuitive. At WYEN, I interviewed him several times and covered his race for governor. Thompson's style earned my respect, and I tried emulating his level of professionalism complemented by an easy manner.

I thought about attending one final meeting on radio writing but decided to head back to campus. I had only a bicycle, and the ride was an adventure during the day, let alone as the sun went down with me riding on the street competing with cars. So I got up and left the room ... said good-bye to my buddy Mark Woolsey and a few others that I'd see again at graduation. Through the hallway and down the stairs and out the door—that was the idea, but an SIU broadcast professor had started down the stairs behind me. I recognized him immediately. I had to turn around. I couldn't ignore him. He'd been one of the best instructors I had for broadcasting. At the bottom of the staircase, I looked up. I didn't know what to say, but I didn't have to say anything. Professor Robert Kurtz volunteered advice.

"Once you turn on a microphone, it'll be most difficult to ever turn it off."

"Thank you," I said, somewhat unsure of the meaning. I appreciated his wisdom. I never forgot his words because I figured someday I'd put it to a test. About thirty years later, as I considered whether to continue working in radio, I fully unwrapped the meaning and recognized Kurtz was right. Turning off the microphone was not possible for me.

Using my name, Kurtz wished me luck in my career. I didn't think he'd remember, so many names passed through his classrooms, but he didn't forget. No obligation of classroom teaching forced Kurtz's hand.

The lessons of Thompson and Kurtz weren't too bad to bring with me to WYEN. I also had the whole of Charlie Lynch's SIU broadcasting department. R-TV Chairman Lynch impressed upon me to remain available and reachable even at the highest level of responsibility.

The Walters family gave our news department freedom to pursue stories without them second-guessing our decisions. Editorial freedom meant we'd go after a wide range of stories from covering Gacy to attending the tryouts of future performers at shows for Great America in Gurnee. The announcers were a bit more structured, yet Allen and Mason allowed some creative control within their shift. Smithers had set the direction earlier.

I'd only worked at WYEN for two and a half years before entertaining other career paths. What prompted this change in direction happened early in 1979 when Mr. Walters felt it necessary to part ways. I told him chef school might be something I could find worthwhile, but the words didn't ring true. I couldn't see *turning off the microphone*. But this early in my career, I didn't face the prospect of an internal struggle over leaving broadcasting because I could always return, and I had not truly sacrificed anything—not the physical challenge of rising at 3 a.m. to go to work or putting long hours into producing a ten-minute public affairs program. As I became an "older" veteran broadcaster working around kids in their twenties just out of college, I looked much harder at myself and questioned what I'd want to do with the rest of my career. The professor's words swirled around. Every possible job I thought I could do just didn't have the spark and life of radio, despite the fact I'd have to continue preparing for work at 3 a.m.

Meier, Roberts, Brown, and many of the others of the WYEN family can describe why they keep their microphone turned on after all these years. Although Kurtz didn't tell them how difficult it would be to permanently *switch off*, they are aware as I am. I can only describe what it feels like talking on the radio. Few things in life compare to this feeling. Your thoughts mean something. There is value to what you say because you already know that people want to hear from you. The feedback is in your audience numbers. Also, you can judge value by the activity on your social sites and blogs. As a broadcaster, you can finish your day knowing your ideas passed through lots of filters, mainly your audience. You've made them laugh, cry, or just plain think, and that's something that's hard to give up.

The former WYEN announcers and staff looked back at a time gone by, recognizing themselves as witnesses of something wonderful and life changing. By participating in the history of WYEN Request Radio, this exclusive broadcasting club will never let the Walt-West Enterprises station end up on a website of forgotten call letters.

During the course of writing *The WYEN Experience*, I collected letters from listeners and former WYEN staff.

When I graduated from Maine East High School in 1978, I knew I just had to work in broadcast news. I had caught the bug at the student radio station there, WMTH. A friend's dad, Bob Cronin, worked in sales at WYEN, and I somehow convinced him to slip my tape to News Director Stew Cohen in hopes of a summer internship. I still have Stew's typewritten note—Listened to his tape. He has potential. *It was the thrill of my young lifetime, and the start of a career. For the next couple of months before heading off to college at Wisconsin, I would tool around the Chicago suburbs in my mother's car, doing feature stories that aired on WYEN Metro news on weekends. I would report, write, and produce these stories. This was an opportunity that would be unheard of today for a kid just out of high school. (I think Stew may have somehow thought I was already a college student. Far be it from me to correct him!) I remember doing a story about Frank Lloyd Wright homes in Oak Park, another one about a political commercial archivist at Northwestern University, and profiling a seemingly sane man who tracked UFOs. There was also a woman older than one-hundred-years-old who shared her secret to longevity. I was eighteen and on top of the world. It was my first job (internship) as a reporter. Nearly thirty-five-years later, I'm still at it.* **—Scott Cohn, intern at WYEN, now Senior Correspondent CNBC.**

I worked as a secretary in 1975 and 1976 for WYEN. It felt like a family working for WYEN. For me, this was a great starting point for a career that continued at WDHF/WMET, A&M Records and then back to WMET, WYSP in Philadelphia and WFMT in Chicago, before moving into Cable TV. I really believe none of that would have been possible without the experience and joy that came from that first job in radio. **—Mary Jane (Deasy) Kupsky, secretary at WYEN**

I did weekends at WYEN under Walters and later Kevin Jay and worked in the production department of WEA Chicago (Warner/Electra/Atlanta Records) during the week. I loved my weekends at WYEN. One time I found an old Meier aircheck in the backroom at WYEN and sent it to him when he was with Dahl on the Loop in Chicago. They played snippets of the tape during their entire show one day with Steve making fun of Garry. I ran into Garry one day and admitted that I had sent that, and he thanked me and said it was fun listening to it again. **—Susan Carr, WYEN announcer, now on-air at All News Atlanta's 106.7**

I was a faithful weekend listener. The radio would go on every Saturday and Sunday morning when I woke up and go off when I went to bed around midnight. I worked forty hours a week and attended Northern Illinois University after work. With the commute to DeKalb, I would not get home until 10 p.m. I'd study Saturday for my Monday night and Tuesday night classes and study on Sunday for my Wednesday night and Thursday night classes. The WYEN announcers and Request Radio were my constant companions. The station featured a lot of Barbara Streisand, Carpenters, and the like. One song that was played often was Joni Mitchell's "Big Yellow Taxi" and another staple was Maria Muldaur's "Midnight at the Oasis." My musical tastes were pulling away from the rock and roll of WLS and WCFL. The music of WYEN at that time was "soft rock and adult contemporary." As the decade proceeded, the music began to change, as did the playlist. —**Mike Krickl**

I listened to the station mostly in its early years. I was a student at the University of Illinois in the early 1970s and remember listening to it whenever I'd come home to Chicago on vacation. I remember seeing how close I'd have to get to the city before I could get the station's signal. Once in a rare while, I could pick up the station, albeit with much interference, all the way down in Champaign.

"Request Radio" was a unique niche that played both pop and songs that nobody else played. It made me aware of a lot of different musical genres. I remember hearing that Smithers was the brainchild of the station, but not sure if he selected the music. Later on, the station changed formats a bit and played mostly stuff that everyone else was playing at the time.

The announcer lineup in the early years, to the best of my recollection, was Frank Gray in the early mornings, a guy named Bruce something (Bruce Elliott?) midmornings, Smithers in the afternoons, (replaced by Reynolds), who had a great radio voice, Zur in the evenings, and Peters overnights.

While in high school, I became very interested in radio. I planned on majoring in radio news and even got accepted to Northwestern University's Radio-TV Department, but at the last minute decided that I didn't want to have to start out in a small town in the middle of nowhere, so I changed my mind and went to the U of I and majored in psychology instead. Anyway, one night I was listening to WYEN and called in to request a song. I talked to Peters and asked him if they ever needed any help answering the phones. I figured it would be a good experience and fulfill my fantasy of working at a station. Gil said they didn't but invited me to the station anyway. I went down, I think that very night, and worked into the early morning.

One thing that always amused me was that WYEN reported the "dry weather chances," not the chance of rain. They had great jingles. —**Bruce Bohrer**

I used to listen to WYEN in the late 1970s and loved the music. There was a very nice song played about every tenth song by Megan (changed to Megon) McDonough in the late 1970s. It was usually introduced something like "And now here is Chicago's very own Megan McDonough." Do you remember the song? —**Theresa**

Three McDonough songs WYEN played were "Do Me Wrong But Do Me" from Sketches *album*, "Daddy Always Liked a Lady" and "Wishing for You" from the Keepsake *album*. —**Rob Reynolds**

For all the people who wrote about their connection to WYEN, thank you!

In the Preface of *The WYEN Experience,* Alice fell into a hole and found a whole world of entertaining characters in Wonderland. WYEN staff found a Wonderland too, down the steps of the office building, through the station door, and into a world of wonderful WYEN characters described in this book. That's where Tanner waited to tell you his story ... and Buckley, his story ... and Smithers, his story ... and Stouffer, her story ... and the staff list goes on and on; no Cheshire Cats, Mad Hatters, March Hares, or Queen of Hearts behind our door.

CAROL AND ED JOIN ACTRESS SALLY KELLERMAN FOR A PICTURE.

In my imagination, Mr. Walters gave me a hug for bringing back to life the radio station of his dreams. He'd have hugged Elliott, Reynolds, Devine, Neches, and the rest of the WYEN family for keeping alive the very thing he enjoyed most in his life. We talked about radio ... about Walters's WYEN. He and Carol have earned a place in our hearts and minds.

Make the Most of Music Magic
Hear it Clear Across the Lake
Catch our Waves in Stereo
We've got Musical Things to Say

O'Hare Airport, the CTA
From Milwaukee to Gary
Listen to Request Radio
WYEN

Thank you, Mr. Ed Walters and Mrs. Carol Walters!

CPSIA information can be obtained at www.ICGtesting.com
Printed in the USA
LVOW130151160313

324535LV00001BA/1/P